Topology in Optics (Second Edition)

Tying light in knots

Topology in Optics (Second Edition)

Tying light in knots

David S Simon
Department of Physics and Astronomy, Stonehill College, Easton, MA, USA

IOP Publishing, Bristol, UK

ISBN 978-0-7503-3471-6 (ebook)
ISBN 978-0-7503-3469-3 (print)
ISBN 978-0-7503-3472-3 (myPrint)
ISBN 978-0-7503-3470-9 (mobi)

DOI 10.1088/978-0-7503-3471-6

Version: 20210501

IOP ebooks

British Library Cataloguing-in-Publication Data: A catalogue record for this book is available from the British Library.

Published by IOP Publishing, wholly owned by The Institute of Physics, London

IOP Publishing, Temple Circus, Temple Way, Bristol, BS1 6HG, UK

US Office: IOP Publishing, Inc., 190 North Independence Mall West, Suite 601, Philadelphia, PA 19106, USA

Dedicated to Marcia and Alee, and of course, the cats.

Contents

Preface

Topology is the study of properties of geometrical objects that remain invariant as the object is bent, twisted, or otherwise continuously deformed. It has been an indispensable tool in particle physics and solid-state physics for decades, but in recent years, it has become increasingly relevant in classical and quantum optics as well. It makes appearances through such diverse phenomena as Pancharatnam–Berry phases, optical vortices and solitons, and optical simulations of solid-state topological phenomena.

The goal of this book is to provide in concise form the necessary mathematical background needed to understand these developments and to give a rapid survey of some of the optical applications where topological issues arise. The level of presentation should make it accessible to advanced undergraduates in mathematics, physics, and related areas. Needless to say, the treatment of these topics is far from exhaustive, but it is hoped that this book this will whet the appetite of the reader and lead him or her to learn about these topics in more detail via the original literature.

David Simon
Boston, MA

Acknowledgements

I would like to thank my friends and colleagues at Stonehill College and Boston University for their all their help and support over the years, including Shuto Osawa and Zach Furman, as well as professors, Guiru Gu, Gregg Jaeger, Alessandro Massarotti, and Alexander Sergienko, and the late Michael Horne. Thanks also to the very helpful people at IOP Publishing, especially Ashley Gasque and Robbie Trevelyan.

Author biography

David Simon

David Simon received a bachelor's degree in mathematics and physics from Ohio State University, followed by doctoral degrees in theoretical physics (Johns Hopkins) and engineering (Boston University). Originally trained in mathematical physics and quantum field theory, he now works primarily in quantum optics and related areas. He has been the author or coauthor of dozens of papers on topics ranging from the use of supersymmetry in quantum mechanics to the application of quantum entanglement to optical measurement and cryptography. After more than a decade teaching at Nova Southeastern University in Fort Lauderdale, he is currently Professor of Physics in the Department of Physics and Astronomy at Stonehill College (Easton, MA) and a visiting researcher at Boston University.

IOP Publishing

Topology in Optics (Second Edition)
Tying light in knots
David S Simon

Chapter 1

Topology and physics: a historical overview

1.1 Introduction: searching for holes in fields of light

Light is a particular type of excitation of the electromagnetic field, one in which the electric and magnetic fields oscillate in phase with each other. Recall that, from a mathematical point of view, vector fields (including the electric or magnetic fields) are simply spatially varying collections of vectors, with one vector assigned to each point in space. These fields normally vary continuously and differentiably as functions of position. But in three dimensions, there can be isolated points or curves on which this well-behaved nature breaks down. These *singular points* or *singular curves* are subsets of space on which some variable associated with the field (such as phase or polarization) is undefined; since the field properties must be well-defined at each point, this forces the amplitude of the field to vanish at that point. In analogy to fluid mechanics, such singular points or curves are often called *vortices* or *vortex lines*, and they appear in many optical contexts, such as laser speckle, orbital angular momentum-bearing 'donut' or 'corkscrew' beams, and light beams with a variety of other sorts of nontrivial spatially dependent structure.

It has long been known that topology, the study of geometric properties that remain invariant under continuous deformations of a space, is useful for studying and classifying systems with singularities. The set of possible field configurations can be viewed as an abstract space, and the singularities amount to holes in this space. The normal strategy for studying a space with holes is to look at collections of spheres of different dimensions and see whether or not they can be contracted to a point without crossing any holes. This technique is called *homotopy theory* and is one of the major pillars of algebraic topology. Homotopy theory will be discussed in some detail in chapter 3.

We will see later that quite complex singularity structures can appear in optics, including multiple optical vortex lines that can form complicated knots and that can become linked or braided with each other.

doi:10.1088/978-0-7503-3471-6ch1

Topology can enter optics in many other ways, beyond the study of vortices. For example:

- A different type of singularity may occur when wavefronts 'pile-up' on top of each other, leading to a sudden amplitude change when some curve is crossed. These ray-optical discontinuities are called *caustics* [1].
- Topology also enters optics via the study of non-dynamical phases (*Berry phases, Pancharatnam phases*, etc). These are phase factors acquired by a quantum mechanical wavefunction that are not due to the usual Hamiltonian time evolution factor e^{-iHt}. They are instead a result of changes in the parameters that define the Hamiltonian. In the optical case, these parameters might include, for example, the index of refraction or the birefringence of a material. Non-dynamical phases of this type have made appearances in nearly every area of physics over the past 30 years and can be of either geometrical or topological origin. It will be seen in chapter 9 that such phases arise in optical systems in a variety of ways.
- When the fields at different boundary regions of a space are in states with different topological properties, the field solution can give rise to localized, particle-like wave pulses called *solitons*. These solitons occur in fields ranging from fluid dynamics to elementary particle physics, and optical solitons have become an important topic in fiber optic systems.
- One further area where topology has entered optics in just the past few years is in the construction of optical systems that can mimic the behavior of unusual solid-state systems known as topological insulators. Topological insulators are systems with periodic lattice structures that are insulators in their interior (their bulk), but which act as conductors on their boundaries. Such materials have associated topological invariants and are governed by Hamiltonians that wrap in nontrivial ways around some compact space as a set of parameters is varied. Such systems have unusual properties that will be discussed in chapter 10, and several types of optical systems have been constructed or proposed in which photons simulate the unusual topological behavior of their solid-state analogs. A survey of these optical analogs will be given in chapter 11.

All of the areas discussed above contain some features in common. In particular, each of the systems has solutions that are characterized by various integer-valued quantities, generically called *topological quantum numbers*. The inability of these discrete numbers to vary continuously leads to greater stability of the system's properties than might otherwise be expected. The essence of such topological quantum numbers is that they measure *global* properties that belong to the system as a whole, rather than local properties like temperature or pressure that can be determined by measurements at individual points.

Topology has become ubiquitous in physics, making prominent appearances in subjects ranging from solid-state physics to superstring theory. In the remainder of this chapter, we give a brief historical overview of the rise of topological methods in physics.

1.2 Topology and physics

Although earlier moments, such as Euler's use of graph theory to investigate the Konigsberg bridges problem (1736) could be singled out as the beginning of the study of topology, the subject only really became an important area of mathematics with the work of Poincaré in the 1880s and 1890s. While investigating the properties of solutions to differential equations, and especially problems in celestial mechanics, he was led to the study of smooth mappings between surfaces, to fixed points, singularities of vector fields, and other topics that would now be considered topological. He went on to give the first definitions of homotopy and homology and to lay the foundations of modern algebraic topology. Poincaré's topological studies of solutions to differential equations as curves on manifolds was continued in the early 20th century by Birkhoff and others, with the results eventually being systematically applied to mechanical systems by Kolmogorov, Arnold, and Moser. Simultaneously, other branches of the subject, such as differential topology and combinatorial topology began expanding, leading to a number of fixed point theorems and to the clarification of useful concepts such as compactness, connectedness, and dimension.

Aspects of topology, then known as *analysis situs* or *geometria situs*, had made appearances in physics before this, of course. For example, Gauss' law and Ampère's law in electrodynamics are both topological in nature: they involve line or surface integrals that remain invariant under continuous deformations of the underlying curve or surface; in modern terminology, we would say that these integrals (the electric and magnetic fluxes) are topological invariants. In fact, integer linking numbers (chapter 5) made their first appearance in a study by Gauss of Ampère's law.

Similarly, in fluid mechanics the study of vortices has a long history. Then, starting in the 1860s, Peter Tait and William Thomson (Lord Kelvin) tried to model atoms as knotted vortex lines in the ether. The motivations included the fact that the multiplicity of different atoms could be explained by the variety of different ways a vortex line could be knotted, and the fact that the stability of atoms could be attributed to the inability to untie a knot without cutting it open; in other words, atomic stability follows from topological stability of the knots. Different spectral lines could also be explained by different vibrational modes of the structure. The work of Tait and Kelvin led to knot theory becoming a major branch of topology, but after the idea of a space-filling ether was abandoned, knots disappeared from physics for almost a century, until they re-emerged in superstring theory and statistical mechanics, and then in other areas like optics.

Despite occasional appearances of topological invariants in classical mechanics and electrodynamics, it was not until the advent of quantum mechanics that topology began to gain the central role in theoretical physics that it enjoys today. This key role is largely the result of three developments in quantum mechanics. The first two are Dirac's analysis of magnetic monopoles (1931) and the discovery of the Aharonov–Bohm effect (1959). These are briefly discussed in the next two sections. Much later, the discovery of the quantum Hall effect (see section 10.1) in 1980 led to an explosive expansion of the role of topology in condensed matter physics.

1.2.1 Dirac monopoles

Although never seen experimentally, the possibility of isolated magnetic charges or monopoles has long been studied theoretically, starting with the work of Paul Dirac in the 1930s [2]. In analogy to electric charges, a point-like magnetic monopole should produce a magnetic field (in SI units)

$$\boldsymbol{B} = \frac{\mu_0 g}{4\pi r^2}\hat{r} = -\nabla V(r), \tag{1.1}$$

where g is the magnetic charge and $V = \mu_0 g/4\pi r$ is the magnetic scalar potential. Because of the identity

$$\nabla^2\left(\frac{1}{r}\right) = -4\pi\delta^{(3)}(\boldsymbol{r}), \tag{1.2}$$

the magnetic analog of Gauss' law is

$$\nabla \cdot \boldsymbol{B} = g\mu_0\delta^{(3)}(\boldsymbol{r}), \tag{1.3}$$

where $\delta^{(3)}(\boldsymbol{r})$ is the three-dimensional Dirac delta function and the magnetic charge density is $\rho_m(\boldsymbol{r}) = g\delta^{(3)}(\boldsymbol{r})$.

Recall that when a particle of momentum $\boldsymbol{p}$ propagates with displacement $\boldsymbol{r}$, the wavefunction picks up a phase factor,

$$\psi \to \psi e^{i\boldsymbol{p}\cdot\boldsymbol{r}/\hbar} \tag{1.4}$$

The phase of a single wavefunction at a given point has no physical relevance, but the phase *difference* between points is meaningful, since it is measurable through interference effects. When there is a field present, the minimal coupling procedure of electromagnetism leads (for a particle of charge e) to an effective shifting of the momentum,

$$\boldsymbol{p} \to \boldsymbol{p} - \frac{e}{c}A. \tag{1.5}$$

The phase difference between points A and B therefore gains a path-dependent contribution

$$\Delta\phi = \frac{e}{\hbar c}\int_A^B \boldsymbol{A}(\boldsymbol{r}) \cdot d\boldsymbol{l}, \tag{1.6}$$

where $d\boldsymbol{l}$ is the length element tangent to the integration path.

Now consider a closed loop C. The electric and magnetic fluxes through the loop are related to the electric and magnetic charges e and g by Gauss' and Ampère's laws:

$$\Phi_E = \int_S \boldsymbol{E} \cdot d\boldsymbol{s} = \frac{e}{\epsilon_0} \qquad \Phi_B = \int_S \boldsymbol{B} \cdot d\boldsymbol{s} = \mu_0 g, \tag{1.7}$$

where S is any surface bounded by C. The phase change when the particle is transported around this loop is proportional to the magnetic flux enclosed by C, Φ_B:

$$\Delta\phi = \frac{e}{\hbar c}\oint_C \boldsymbol{A}(\boldsymbol{r})\cdot d\boldsymbol{l} = \frac{e}{\hbar c}\int_S \boldsymbol{B}\cdot d\boldsymbol{s} = \frac{e}{\hbar c}\Phi_B, \tag{1.8}$$

where $d\boldsymbol{s}$ is the area element, S is a smooth, oriented surface enclosed by C, and Stokes' theorem was used in the second equality. But the wavefunction at the initial point must be single valued, which implies that the phase shift must be an integer multiple of 2π:

$$\Delta\phi = 2\pi n. \tag{1.9}$$

Using equations (1.7) and (1.8), this in turn implies that the magnetic flux is quantized:

$$\Phi_B = \frac{2\pi\hbar c}{e}n = \frac{hc}{e}n. \tag{1.10}$$

In fact, Dirac showed that the product of the electric and magnetic charges must also be quantized:

$$eg = \left(\frac{hc}{\mu_0}\right)n, \tag{1.11}$$

for integer n. Thus, the existence of magnetic monopoles would explain the quantization of electric charge that is actually seen in nature.

The integer n is found to be a topological quantum number, counting the windings of mappings around the singular point where the monopole resides. In modern terms, the Dirac quantization condition states that the electromagnetic field tensor integrated over a two-dimensional sphere S^2 enclosing the monopole is an integer:

$$c_1 = \int_{S^2} \frac{F_{\mu\nu}}{2\pi} dx^\mu \wedge dx^\nu = n. \tag{1.12}$$

The field tensor $F_{\mu\nu}$ plays the role of a curvature on a fiber bundle. The resulting integer is called the first Chern number and will be discussed in more detail in chapter 5.

The origin of this quantization is the singular point in the field at $r = 0$. Dirac in fact found a singular vortex line along the negative z-axis. This vortex line is unphysical and its location can be moved around by gauge transformations, but there is no single gauge choice defined at all points in space that will eliminate the vortex everywhere. However, Wu and Yang [3, 4] showed that the vortex line could be eliminated by defining two overlapping coordinate regions; the gauge fields on the overlap region are related by an appropriate gauge transformation. The work of Wu and Yang illustrated the fact that the proper mathematical setting for the description of electromagnetism (and, more generally, of all gauge theories) is the theory of fiber bundles (chapter 4).

Studies of topological structures in quantum field theories took off in the 1970s, beginning with the discovery of quantized magnetic flux lines in superconductors [5], which (taken together with Dirac's work) led to the realization that magnetic monopoles should exist in non-Abelian field theories [6, 7].

It has now been found that monopoles occur in many quantum field theory models and are ubiquitous in grand unified field theories in particle physics. Magnetic monopole-like structures can also occur in other types of physical systems (see chapter 10, for example).

1.2.2 Aharanov–Bohm effect

A second demonstration of the importance of topology in physics came with the Aharonov–Bohm effect [8]. Consider a charged particle moving in the vicinity of a current carrying solenoid. There is a magnetic field $\boldsymbol{B} \neq 0$ inside the solenoid, but the field vanishes outside. The vector potential, $\boldsymbol{A}$, however is nonzero everywhere, inside and out. Prior to the rise of the Aharonov–Bohm effect, it was believed that the field $\boldsymbol{B}$ was the physically important variable and that $\boldsymbol{A}$ was simply a mathematical convenience of no physical significance. However, Aharonov and Bohm showed that when the particle circles the solenoid in a closed loop C, staying entirely in the $\boldsymbol{B} = 0$ region, there is nevertheless a phase shift given by

$$\Delta\phi = \frac{e}{\hbar c}\int_C \boldsymbol{A}(\boldsymbol{r}) \cdot d\boldsymbol{l} = \frac{e}{\hbar c}\int_S \boldsymbol{B} \cdot d\boldsymbol{s}, \tag{1.13}$$

and that this shift is an integer multiple of 2π. The existence of the Aharonov–Bohm effect was verified in an experiment by Chambers in 1960 [9].

The solenoid contains a singularity in the vector potential. One can therefore view the solenoid as a hole in the space of allowed field configurations. The quantization arises from the topological fact that curves in $\boldsymbol{A}$-space that enclose the solenoid are non-contractible. The integer n here counts the number of times the loop encloses the singularity: it is a winding number. This winding number characterizes the distinct homotopy classes (see chapters 3 and 5) of the field. The Aharanov–Bohm phase accumulated as the electron circles the solenoid is an example of the geometric Berry phase to be discussed in chapter 9.

1.2.3 Topology in optics

By the 1990s and 2000s, many of the topology-related structures previously found in other areas of physics began to come up in optics. For example, vortices and vortex lines, winding numbers and linking numbers, and even non-orientable Möbius strips have all made appearances in various areas of optics. Further, the Aharonov–Bohm effect is a special case of the geometric or Berry phase; the first known description of a geometric phase appeared in a study of polarization optics in the 1950s, although its significance was not widely recognized for decades.

All of these topics will be described in coming chapters. The range of optical phenomena in which topology plays a role has become large, so in a book of this size some of them will necessarily be treated only in the briefest of terms, but hopefully enough of a flavor will be given to interest the reader in pursuing a deeper study via the provided references.

As general references to the broader background material, we list a few useful texts here. Many excellent introductions to algebraic and differential topology may be found, including [10–15]. Numerous reviews covering applications of topology to gauge field theory, particle physics, and condensed matter physics also exist, which physicists and engineers may find more accessible; these include [16–20]. The history of topology and of its applications in physics are reviewed in [21] and [22], respectively.

References

[1] Nye J 1999 *Natural Focusing and Fine Structure of Light: Caustics and Wave Dislocations* (Boca Raton, FL: CRC Press)
[2] Dirac P A M 1931 *Proc. Roy. Soc. London* **A133** 60
[3] Wu T T and Yang C N 1975 *Phys. Rev.* D **12** 3843
[4] Wu T T and Yang C N 1975 *Phys. Rev.* D **12** 3845
[5] Nielsen H and Oleson P 1973 *Nucl. Phys.* B **61** 45
[6] t'Hooft G 1974 *Nucl. Phys.* B **79** 276
[7] Polyakov A M 1974 *JETP Lett.* **20** 194
[8] Aharonov Y and Bohm D 1959 *Phys. Rev.* **115** 485
[9] Chambers R G 1960 *Phys. Rev. Lett.* **5** 3
[10] Hatcher A 2002 *Algebraic Topology* (Cambridge: Cambridge University Press)
[11] Greenberg M and Harper J 2018 *Algebraic Topology: A First Course* (Boca Raton, FL: CRC Press)
[12] Munkres J R 2000 *Topology* (Englewood Cliffs, NJ: Prentice-Hall)
[13] Basener W F 2006 *Topology and Its Applications* (Hoboken, NJ: Wiley)
[14] Hirsch M W 1997 *Differential Topology* (Berlin: Springer)
[15] Guillemin V and Pollock A 1974 *Differential Topology* (Englewood Cliffs, NJ: Prentice-Hall)
[16] Moriyasu K 1983 *An Elementary Primer for Gauge Theory* (Singapore: World Scientific)
[17] Morandi G 1992 *The Role of Topology in Classical and Quantum Physics* (Berlin: Springer)
[18] Nakahara M 2003 *Geometry, Topology and Physics* 2nd edn (Boca Raton, FL: CRC Press)
[19] Nash C 1992 *Differential Topology and Quantum Field Theory* (London: Academic)
[20] Nash C and Sen S 2013 *Topology and Geometry for Physicists* (Mineola, NY: Dover)
[21] James I M 1999 *History of Topology* (Amsterdam: Elsevier)
[22] Nash C 1997 *Topology and Physics–A Historical Essay*, arXiv:hep-th/9709135

IOP Publishing

Topology in Optics (Second Edition)

Tying light in knots

David S Simon

Chapter 2

Electromagnetism and optics

2.1 Electromagnetic fields

In the 1860s, the Scottish physicist James Clerk Maxwell gathered together all that was known at the time about electricity and magnetism and showed that it all followed from a small set of equations now known as the Maxwell equations. In modern vector notation and SI units, the differential form of these laws is given by

$$\nabla \cdot \boldsymbol{E} = \frac{\rho}{\epsilon_0}, \qquad \nabla \cdot \boldsymbol{B} = 0 \tag{2.1}$$

$$\nabla \times \boldsymbol{E} = -\frac{\partial \boldsymbol{B}}{\partial t}, \quad \nabla \times \boldsymbol{B} = \mu_0\left(J + \epsilon_0 \frac{\partial \boldsymbol{E}}{\partial t}\right) \tag{2.2}$$

The electric and magnetic fields can be computed from the scalar potential $\phi(\boldsymbol{r}, t)$ and the vector potential $\boldsymbol{A}(\boldsymbol{r}, t)$,

$$\boldsymbol{B} = \nabla \times \boldsymbol{A} \qquad \text{and} \qquad \boldsymbol{E} = -\nabla\phi - \frac{\partial \boldsymbol{A}}{dt}. \tag{2.3}$$

The potentials ϕ and $\boldsymbol{A}$ are not entirely well-defined: for any function $f(\boldsymbol{r}, t)$, the **gauge transformations**

$$\boldsymbol{A} \to \boldsymbol{A} + \nabla f \qquad \text{and} \qquad \phi \to \phi - \frac{\partial f}{\partial t} \tag{2.4}$$

leave $\boldsymbol{E}$ and $\boldsymbol{B}$ unchanged, along with all other physically measurable quantities. This ambiguity in the potentials is sometimes useful, since it can often be utilized to put them into a form that simplifies a given problem. However, it also introduces conceptual difficulties and raises the question of whether the potentials are physically 'real' in the same way the directly measurable $\boldsymbol{E}$ and $\boldsymbol{B}$ fields are. We will see later that the gauge invariance in fact has geometric and topological meaning.

doi:10.1088/978-0-7503-3471-6ch2

Under Lorentz transformations, the electric and magnetic fields become linear combinations of each other, so in relativistic theories it is natural to view them as components of a single second-rank electromagnetic field tensor,

$$\mathcal{F} = \begin{pmatrix} 0 & E_x/c & E_y/c & E_x/c \\ -E_x/c & 0 & -B_z & B_y \\ -E_y/c & B_z & 0 & B_x \\ -E_z/c & -B_y & B_x & 0 \end{pmatrix}. \tag{2.5}$$

The components of the field tensor can be more concisely written as

$$F_{\mu\nu} = \partial_\mu A_\nu - \partial_\mu A_\nu, \tag{2.6}$$

where μ and ν are space–time indices, running from 0 (for time) to 3 (with 1 to 3 representing space directions). Here, A_μ is a four-vector with ϕ and (A_x, A_y, A_z), respectively, as its time and space components, while c is the speed of light in vacuum. A_μ is the **gauge potential** or **gauge field**. The contraction of $F_{\mu\nu}$ with itself, $F_{\mu\nu}F^{\mu\nu}$, is a Lorentz invariant and in fact is proportional to the Lagrangian density of the electromagnetic field. (Here, we are using the Einstein summation convention, where repeated indices, one up and one down, are summed over so that $F_{\mu\nu}F^{\mu\nu}$ is short for $\sum_{\mu\nu} F_{\mu\nu}F^{\mu\nu}$.)

Combining the various Maxwell equations and using standard vector identities, it is readily shown that $\boldsymbol{E}$ and $\boldsymbol{B}$ obey a pair of wave equations,

$$\left(\nabla^2 - \frac{1}{c^2}\frac{\partial^2}{\partial t^2}\right)E_j(\boldsymbol{r}, t) = 0 \tag{2.7}$$

$$\left(\nabla^2 - \frac{1}{c^2}\frac{\partial^2}{\partial t^2}\right)B_j(\boldsymbol{r}, t) = 0, \tag{2.8}$$

where $j = x, y, z$ labels the spatial components, and

$$\nabla^2 = \frac{\partial^2}{\partial x^2} + \frac{\partial^2}{\partial y^2} + \frac{\partial^2}{\partial z^2} \tag{2.9}$$

is the Laplacian. Maxwell's wave equations describe the propagation of transverse electromagnetic waves through vacuum at the speed

$$c = \frac{1}{\sqrt{\epsilon_0 \mu_0}} \approx 299\,792\,458 \text{ m s}^{-1} \approx 3 \times 10^8 \text{ m s}^{-1}. \tag{2.10}$$

Inside matter, the speed is reduced to $\mathrm{v} = \frac{c}{n}$, where n is the refractive index of the material. Since $E(\boldsymbol{r}, t)$ and $B(\boldsymbol{r}, t)$ are proportional to each other for electromagnetic waves, it is often sufficient to consider just the electric field to understand the wave's behavior. In coming chapters, we will often follow standard practice and allow $\boldsymbol{E}(\boldsymbol{r}, t)$ to be complex; the physical electric field is then given by its real part. Such a

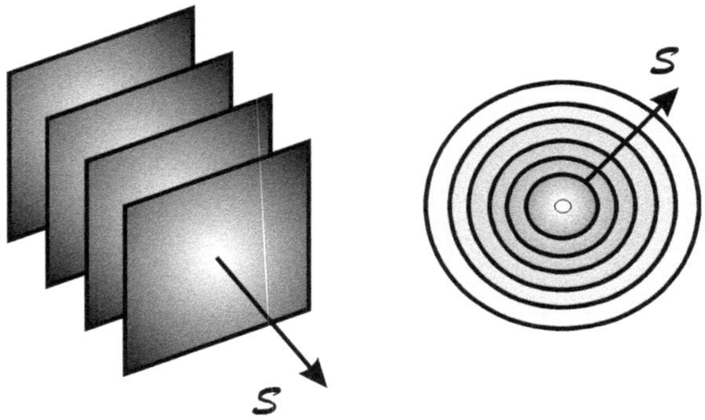

Figure 2.1. The wavefronts for a plane wave (left) are parallel planes, with Poynting vector $\mathcal{S}$ perpendicular to the planes and parallel to the propagation direction. For spherical waves (shown in cross-section on the right) the wavefronts are nested spheres, either expanding outward or collapsing inward from a point source at the center. The Poynting vector is pointing radially outward (for an expanding wave) or radially inward (for a contracting wave).

practice is especially convenient in quantum theory, and makes effects such as interference and nonlinear processes easier to treat.

Recall that wavefronts are surfaces of constant phase. The Poynting vector, $\boldsymbol{S} = (1/\mu_0)\boldsymbol{E} \times \boldsymbol{B}$, which describes the flow of energy, is always perpendicular to these surfaces. The archetypal solution to the electromagnetic wave equation is the plane wave (figure 2.1, left), in which the wavefronts are parallel planes of infinite extent, perpendicular to the propagation vector of the wave. The electric and magnetic fields are perpendicular to each other and to the Poynting vector. In reality, there are no exact plane waves, due to their infinite spatial extent and the infinite energy they carry, but they are useful approximations when the wavefronts are sufficiently flat, in the sense that their radii of curvature are large compared to all other relevant size scales in the problem.

The Maxwell wave equations have many other solutions. For example, point sources of light produce spherical waves that propagate outward from the source in all directions (figure 2.1, right). Other, more complicated sorts of waves are possible, including those of highly directional beams, in which the light travels primarily along a particular axis (the z-axis, referred to as the **longitudinal** direction). The electric and magnetic fields of transverse waves will then be in the x–y or **transverse plane**. For beam-like solutions, it is useful to simplify the Maxwell wave equation down to the **Helmholtz equation**. Let ω be the angular frequency of the wave, and then separate the time dependence of the field off from the space dependence: $\boldsymbol{E}(\boldsymbol{r}, t) = \boldsymbol{A}(\boldsymbol{r})e^{-i\omega t}$. Substituting this into the wave equation, we arrive at the Helmholtz equation,

$$(\nabla^2 + k^2)A_j(\boldsymbol{r}) = 0. \tag{2.11}$$

The magnitude of the wavevector $\boldsymbol{k}$ is given by the wavenumber $k = 2\pi/\lambda = \omega/c$. The Helmholtz equation is the usual starting point for studying any type of directed electromagnetic wave motion.

In the quantum mechanical view of the world, light is formed from a stream of particle-like quanta or excitations of the electromagnetic field, called **photons**. These photons have momentum proportional to the wavevector, $\boldsymbol{p} = \hbar \boldsymbol{k}$ and the momentum in turn is the eigenvalue of a differential operator, $\hat{p} = -i\hbar\nabla$. Similarly, the energy $E = \hbar\omega$ is the eigenvalue of the Hamiltonian operator, $\hat{H} = i\hbar(\partial/\partial t)$ and is directly proportional to the frequency.

In the following, the beam axis will always be taken to be the z-axis, and we will often use cylindrical coordinates (r, z, ϕ), where ϕ (the **azimuthal** angle) is the angle about the z-axis. The two-dimensional coordinate vector in the x–y plane (the **transverse** plane) will be denoted $\boldsymbol{r}_\perp$. The z-direction will be referred to as the **longitudinal** direction. The radial or transverse wavevector component $k_r = \sqrt{k_x^2 + k_y^2}$ and the longitudinal component k_z must obey $k_r^2 + k_z^2 = k^2$.

Many types of beamlike solution are possible [1]. Here we simply state the form of the simplest possibility, a **Gaussian beam** [1, 2], in order to define some beam parameters that will come up later. The beam's amplitude profile in any transverse plane will be a Gaussian function. The width of this Gaussian initially decreases as a function of z after leaving the light source (a laser), reaching a minimum at a point called the **waist**, then it slowly increases again (figure 2.2). We take $z = 0$ to be at the waist. The field E and intensity I of the Gaussian beam are given by

$$E(\boldsymbol{r}_\perp, z) = \sqrt{I_0}\,\frac{w_0}{w(z)} e^{-r^2/w^2(z)} e^{ikz + ikr^2/2R(z) - i\zeta(z)}, \tag{2.12}$$

$$I(\boldsymbol{r}, z) = |E(\boldsymbol{r}_\perp, z)|^2 = I_0\left(\frac{w_0}{w(z)}\right)^2 e^{-2r^2/w^2(z)}. \tag{2.13}$$

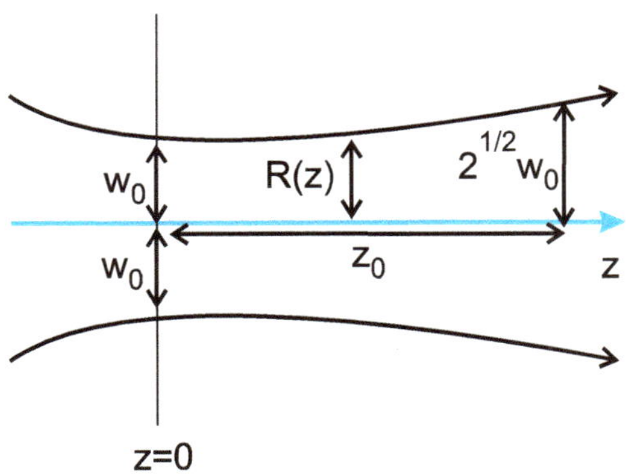

Figure 2.2. The envelope of a Gaussian beam. The light source is at the left edge, with the beam propagating to the right. The radius $R(z)$ as a function of horizontal position z reaches a minimum value of w_0 at the waist, $z = 0$, then expands. The Rayleigh range z_0 is the distance from the waist at which the intensity drops to half its maximum value. At z_0 the radius is $R(z_0) = \sqrt{2}\,w_0$.

Here, the beam radius at distance z is

$$w(z) = w_0\left[1 + \left(\frac{z}{z_0}\right)^2\right]^{1/2}, \tag{2.14}$$

and the radius of curvature of the wavefront is

$$R(z) = z\left[1 + \left(\frac{z_0}{z}\right)^2\right]. \tag{2.15}$$

The factor

$$\zeta(z) = \tan^{-1}\left(\frac{z}{z_0}\right) \tag{2.16}$$

is the **Guoy phase**, which varies from $-\pi/2$ to $+\pi/2$ as the beam propagates from $z = -\infty$ to $z = +\infty$. The maximum intensity I_0 occurs on-axis at $z = 0$. The **Rayleigh range** z_0 is the distance at which the beam cross-section is twice as large as at the waist, or equivalently the distance at which the on-axis intensity is half of its value at the waist: $I(z_0) = (1/2)I_0$. The Rayleigh range and the waist radius are related by

$$w_0 = \sqrt{\lambda z_0/\pi}. \tag{2.17}$$

Note that a more highly focused beam (smaller w_0) will also diverge more rapidly (smaller z_0) as one moves along the z-axis.

2.2 Electromagnetic potentials and gauge invariance

Returning to the gauge potential A_μ defined above, the effect of the electromagnetic field acting on a particle of charge q may be introduced via the minimal coupling principle, replacing the free-particle four momentum $p_\mu = -i\hbar\partial/\partial x_\mu$ by the canonical momentum

$$p_\mu = -i\hbar(\partial/\partial x_\mu) + qA_\mu. \tag{2.18}$$

Here we are using relativistic four-vector notation, where $\mu = 0$ corresponds to the time-like component and $\mu = 1, 2, 3$ are the space-like components:

$$A_\mu = \{\phi, \boldsymbol{A}\} \tag{2.19}$$

$$\partial_\mu = \frac{\partial}{\partial x_\mu} = \left\{\frac{\partial}{\partial t}, \nabla\right\} \tag{2.20}$$

$$p_\mu = \{E, \boldsymbol{p}\}. \tag{2.21}$$

The Schrödinger equation is then of the form

$$\left(\frac{1}{2m}(i\hbar\nabla + q\boldsymbol{A})^2 + q\phi\right)\psi(\boldsymbol{x}, t) = i\hbar\frac{\partial}{\partial t}\psi(x, t) \tag{2.22}$$

Note that the transition to the canonical momentum may also be viewed as starting from the field-free Schrödinger equation,

$$\frac{\hbar^2}{2m}\nabla^2\psi(\boldsymbol{x}, t) = i\hbar\frac{\partial}{\partial t}\psi(x, t), \tag{2.23}$$

and replacing the ordinary derivatives $\partial_\mu \equiv \partial/\partial x_\mu$ ($\mu = 0, 1, 2, 3$) by the covariant derivatives

$$D_\mu = \partial_\mu + \frac{iq}{\hbar}A_\mu. \tag{2.24}$$

The covariant derivative and the gauge potential have clear geometric meanings, as will be seen in chapter 4. Because of this geometric interpretation of the gauge potentials, the A_μ are often referred to as **gauge connections**, since they provide a means of comparing vectors at different points on a curved space via a path connecting those points. The gauge connections here play the same role in electromagnetism that the **Christoffel symbols** or **gravitational connection coefficients**, $\Gamma^\mu_{\nu\lambda}$, play in general relativity. In chapter 9, we will also see that quantities formally identical to gauge connections can appear in other systems, which will lead to the important topic of geometric and topological phases.

The Schrödinger equation and its relativistic generalizations are invariant under the **gauge transformation**

$$\psi \rightarrow e^{\frac{iq}{\hbar}\lambda(x)}\psi, \qquad A_\mu \rightarrow A_\mu - \partial_\mu\lambda(x), \tag{2.25}$$

and a choice of function $\lambda(x)$ is called a choice of gauge. Under this transformation it is readily verified that the wavefunction transforms covariantly, or in other words, that the covariant derivative of ψ transforms the same way as ψ itself:

$$D_\mu\psi \rightarrow e^{\frac{iq}{\hbar}\lambda}D_\mu\psi. \tag{2.26}$$

The idea of gauge invariance is one of the foundational principles of modern physics. All of the fundamental forces in Nature (electromagnetism, gravity, and the nuclear forces) are invariant under some generalization of this transformation. The requirement of gauge invariance strongly constrains the forms of the forces that can exist in Nature, and leads to the existence of conserved quantities. In the case of electromagnetism, the conserved quantity is simply electric charge. The gauge invariance is required to be local, in the sense that the phase of the wavefunction can be chosen independently at each point in space; in other words, $\lambda(x)$ can be an arbitrary function of position, and not restricted to being a constant. Maintaining this invariance requires the introduction of the gauge field A_μ, and it forces A_μ to transform as in equation (2.25) under gauge transformations, so that the gauge ambiguities of the two field A_μ and ψ cancel each other in the Lagrangian and equations of motion. Invariance under more generalized types of gauge transformations leads to the introduction of nuclear forces and gravitational fields.

Gauge transformations are similar to changes of coordinate system. The gauge choice used may affect the intermediate details of a calculation, and some gauge choices may make a particular calculation simpler; however, the choice of gauge should not affect the value of physical observables. Therefore, all quantities that are measurable, such as energy or angular momentum, are required to be gauge-invariant. It can be seen that the gauge potential is not a directly measurable quantity, since its value can be altered by a gauge transformation. However, the electromagnetic field tensor $\mathcal{F}$ of equations 2.5 and 2.6 is gauge-invariant, implying that its components (the electric and magnetic fields) are well-defined.

Let us restrict attention to the case of a purely magnetic field ($\phi = \partial A/\partial t = 0$). Dirac showed that if $\psi_0(\boldsymbol{x}, t)$ is a solution to the Schrodinger equation in the absence of an electromagnetic field, then the solution of equation (2.22) in the presence of the field is given by

$$\psi(\boldsymbol{x}, t) = \psi_0(\boldsymbol{x}, t)e^{\frac{iq}{\hbar}\int \boldsymbol{A}\cdot d\boldsymbol{x}}. \tag{2.27}$$

It should be clear by inspection that if A_μ transforms correctly under gauge transformations (equations (2.25)), then the wavefunction will as well.

Consider the Aharonov–Bohm setup (figure 2.3) [3, 4], in which a charged particle can take either of two paths, one above the region of nonzero field (path C_1) and one below (path C_2). Both paths travel through regions of vanishing magnetic field $\boldsymbol{B}$, but nonzero magnetic potential $\boldsymbol{A}$. The two particles arriving at point Q will differ by a phase:

$$\psi_1 = \psi_0 e^{\frac{iq}{\hbar}\int_{C_1} \boldsymbol{A}\cdot d\boldsymbol{x}} \tag{2.28}$$

$$\psi_2 = \psi_0 e^{\frac{iq}{\hbar}\int_{C_2} \boldsymbol{A}\cdot d\boldsymbol{x}}. \tag{2.29}$$

We may form a closed path C by traveling from P to Q along C_2 and then backwards along C_1 to return to P; in the terminology that will be introduced in chapter 3, C is the product of the other two paths, $C = C_2 * C_1^{-1}$. If the charged

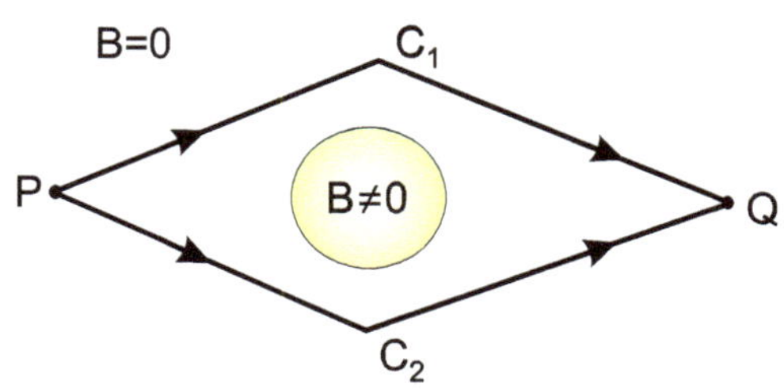

Figure 2.3. Two paths passing a region of nonzero magnetic field (shaded region). The field $\boldsymbol{B}$ is zero along both paths, but the magnetic potential or gauge connection $\boldsymbol{A}$ is nonvanishing.

particle travels around C, its wavefunction as it returns to P may differ from its initial wavefunction by a phase:

$$\psi = \psi_0 e^{\frac{iq}{\hbar}\oint_C \boldsymbol{A}\cdot d\boldsymbol{x}}. \tag{2.30}$$

Using Stokes' theorem and the fact that $\boldsymbol{B} = \nabla \times \boldsymbol{A}$, this can be rewritten as

$$\psi = \psi_0 e^{\frac{iq}{\hbar}\oint_S \boldsymbol{B}\cdot d\boldsymbol{s}} = \psi_0 e^{\frac{iq}{\hbar}\Phi_B}, \tag{2.31}$$

where $\Phi_B = \oint_S \boldsymbol{B} \cdot d\boldsymbol{s} = \oint_C \boldsymbol{A} \cdot d\boldsymbol{x}$ is the magnetic flux through the surface S bounded by C. Notice that the flux only depends on whether or not the curve C encloses the region of nonzero $\boldsymbol{B}$, and how many times it encircles it; otherwise, making continuous deformations of C leaves the both the flux and the resulting phase factor unchanged. In other words, in the present setup with C assumed to be entirely in the region where $\boldsymbol{B} = 0$, the flux through a closed loop is a topological invariant, and the loops can be divided up into equivalence classes called homotopy classes, based on how many times they encircle the $B \neq 0$ region. Homotopy classes will be defined in chapter 3.

One other fact that should be mentioned is that the Dirac phase factors, $e^{\frac{iq}{\hbar}\oint_C \boldsymbol{A}\cdot d\boldsymbol{x}}$ all lie in the unit circle, denoted S^1, on the complex plane. The points on this circle actually form a group (see the appendix for the definition of groups), called $U(1)$, the group of unitary complex numbers. Symmetries or invariances of a physical theory always form groups. $U(1)$ is called the **gauge group** (the local invariance group) of electromagnetism. It is an **Abelian** group, meaning that the elements commute with each other; in contrast, the gauge groups for the nuclear forces and gravity are **non-Abelian** (noncommutative).

2.3 Linear and nonlinear optical materials

In later chapters, some of the effects that will be described will only occur when the light propagates through a material with nonlinear optical response, so in this section we give a very brief review of nonlinear optics. Much more extensive reviews can be found in [5–7].

Consider a linear material first. These are materials in which the electric polarization induced in the molecules by an applied electric field is directly proportional to the field; in our case the electric field will be due to an optical wave passing through the medium. Most materials behave linearly when illuminated with light at low intensities, and any nonlinearity will only become apparent at very high intensity. In linear materials, the index of refraction and the phase velocity are independent of the intensity, and the frequencies of the ingoing and outgoing light waves are equal. Besides being linear, we assume for simplicity that the material in question is nonmagnetic.

For such a linear material, one may define the macroscopic polarization P of a material through the formula familiar from introductory electromagnetism courses,

$$\boldsymbol{P} = \epsilon_0 \chi_1 \boldsymbol{E}, \tag{2.32}$$

where ϵ_0 is the permittivity of free space and χ_1 is the dimensionless linear optical susceptibility of the material. The electric displacement inside the material is

$$\boldsymbol{D} = \epsilon_0 \boldsymbol{E} + \boldsymbol{P} \tag{2.33}$$

$$= \epsilon_0(1 + \chi_1)\boldsymbol{E} \tag{2.34}$$

$$\equiv \epsilon \boldsymbol{E}, \tag{2.35}$$

where the permittivity of the material is

$$\epsilon = \epsilon_r \epsilon_0, \tag{2.36}$$

and the relative permittivity is

$$\epsilon_r = 1 + \chi_1. \tag{2.37}$$

The real part of the relative permittivity determines the index of refraction,

$$n = \sqrt{\mathrm{Re}(\epsilon_r)}, \tag{2.38}$$

while the imaginary part $\mathrm{Im}(\epsilon_r)$ controls the rate of optical absorption in the material. Since we are not concerned here with absorption, we will assume the permittivity is real. The polarization is a linear function of the electric field, and the index of refraction is independent of the magnitude of the optical wave. A passing electromagnetic wave produces a polarization as shown in figure 2.4.

Of course, realistic materials don't behave like this except in some level of approximation. In every material, the index of refraction varies at least slightly with optical intensity. Before the advent of lasers, it was usually safe to ignore these weak nonlinear optical effects. However, lasers can produce very high optical intensities, at which nonlinear effects become strong and can no longer be ignored. Here, we only consider parametric effects, those in the which there is no net exchange of energy between the light and the material (although there may be energy being exchanged between the various different photons within the light beam).

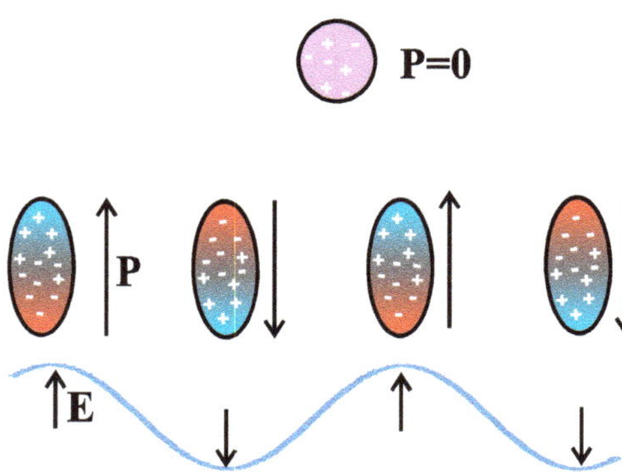

Figure 2.4. Top: unpolarized molecule. Bottom: as an electromagnetic wave passes, the charge in the molecule separates, causing a charge polarization that oscillates as the electric field of the wave alternates in direction.

In the nonlinear case, the polarization need not take the linear form of equation (2.32), but instead can be an arbitrary function of E. We make the simplifying assumptions that the nonlinear response is instantaneous and that we can treat the field as a scalar, in which case $P(E(\boldsymbol{r}, t))$ can be expanded as a power series in $E(\boldsymbol{r}, t)$,

$$P(\boldsymbol{E}) = \epsilon_0(\chi_1 E(\boldsymbol{r}, t) + \chi_2 E^2(\boldsymbol{r}, t) + \chi_3 E^3(\boldsymbol{r}, t) + \cdots) \tag{2.39}$$

$$\equiv P_{lin} + P_{nl}, \tag{2.40}$$

where P_{lin} is the linear part and P_{nl} is the rest. (If we had not assumed an instantaneous material response, the products of $E(\boldsymbol{r}, t)$ at the same time t would instead be convolutions of the fields at different times. Without the scalar approximation, the coefficients χ_j would become tensors.) Physically, a term in P containing a power E^{n-1} describes an interaction with a total of n ingoing and outgoing photons. The χ_2 term is responsible for processes such as parametric upconversion and downconversion, in which two incident photons merge to form a single photon of higher frequency or in which a single incident photon splits to create two outgoing photons of lower frequency. In particular, the χ_2 term is useful for generating pairs of quantum mechanically entangled photons, which provide a basic tool for quantum optics and of modern precision tests of quantum mechanics [8–13].

Here we focus on the case where the χ_2 term can be neglected, so that the cubic term dominates the nonlinear behavior. This occurs in centrosymmetric media, materials where the crystal lattice structure is invariant under the parity transformation, $\boldsymbol{r} \to -\boldsymbol{r}$. In such materials, the symmetry forces the χ_2 term to vanish exactly.

Consider a monochromatic optical wave of frequency ω in such a medium. Denoting the complex electric field by $\mathcal{E}$, it may be written as

$$\mathcal{E}(\boldsymbol{r}, t) = \frac{1}{2}(E(\boldsymbol{r}, t)e^{-i\omega t} + E^*(\boldsymbol{r}, t)e^{i\omega t}). \tag{2.41}$$

Substituting this field into P in place of the original real field, and multiplying out the cube of $\mathcal{E}$, we find terms proportional to $e^{\pm i\omega t}$ and terms proportional to $e^{\pm 3i\omega t}$. We denote these terms by $P(\omega)$ and $P(3\omega)$:

$$P = P(\omega) + P(3\omega). \tag{2.42}$$

These represent outgoing light of frequencies ω and 3ω. Here we are not concerned with the $P(3\omega)$ terms, and focus only on the part where the incident and final frequencies are the same. We can split the $P(\omega)$ part into linear and nonlinear pieces,

$$P(\omega) = P_{lin}(\omega) + P_{nl}(\omega), \tag{2.43}$$

where

$$P_{lin}(\omega) = \epsilon_0 \chi_1 \mathcal{E} \tag{2.44}$$

$$P_{nl}(\omega) = \frac{3}{8}\chi_3\epsilon_0|E|^2\mathcal{E} \equiv \epsilon_0\chi_{nl}\mathcal{E}, \tag{2.45}$$

with $\chi_{nl} = (3/8)\chi_3 I$. Here, $I = |E|^2$ is the intensity. The index of refraction at frequency ω and intensity I can then be written in the form

$$n \approx n_0 + n_2 I, \tag{2.46}$$

where

$$n_0 = \sqrt{1 + \chi_1} \tag{2.47}$$

$$n_2 = \frac{3\chi_3}{8n_0}. \tag{2.48}$$

The most important point to note is that the refractive index is now intensity-dependent. The index determines how the light propagates, but in this case, the spatial distribution of the light also determines how the refractive index varies within the material. So there is a feedback loop set up, in which the light in a sense controls its own propagation. This leads to phenomena such as self-focusing, self-phase modulation, and optically induced transparency, some of which will come up in chapter 8.

Self-focusing occurs when $n_2 > 0$, so that n is larger in regions of higher intensity. In this case a typical light beam, which is most intense near the axis, will bend inward toward the axis; the beam itself causes the material to act like a converging lens. When $n_2 < 0$, the material is called **self-defocusing** and n is larger in regions of lower intensity. We will see in chapter 8 that bright solitons occur in self-focusing materials and dark solitons appear in self-defocusing materials.

With the cubic form of P used in equation (2.45), the material is called a **Kerr medium**, and the dependence of n on intensity is the **optical** or **AC Kerr effect**. It is also sometimes the **quadratic opto-electric effect**, since the interaction energy, proportional to $\boldsymbol{P} \cdot \boldsymbol{E}$, is quadratic in the intensity.

Inserting $E(\boldsymbol{r}, t)$, $P(E)$, and $n(I)$ into the Helmholtz equation and assuming that E is sufficiently slowly varying along the z-axis, the Helmholt equation can be written in the form of a **nonlinear Schrödinger equation** (NLS),

$$\nabla^2 E - \frac{1}{c^2}\frac{\partial^2 E}{\partial t^2} - \frac{1}{\epsilon_0 c^2}\frac{\partial^2 P}{\partial t^2} = 0. \tag{2.49}$$

The nonlinear effects arise from the nonlinear part of the polarization in the last term. Unlike the linear Schrödinger equation in quantum mechanics, the solutions do not obey the superposition principle. Some solutions to the NLS in a Kerr medium will be discussed in chapter 8.

In addition to the Kerr electro-optic effect, there is also a **magneto-optic Kerr effect**, also known as the **Faraday effect**, in which a material induces a phase change in reflected or transmitted light when a magnetic field is applied. This effect comes up in attempts to induce topological effects in photonic systems (chapter 11), because it provides a way to break time-reversal symmetry. A material in which magneto-optic effects are prominent is sometimes called a **gyrotropic material**.

2.4 Polarization and the Poincaré sphere

Recall that light is said to be **linearly polarized** if its electric field vector remains in a fixed direction, called the **polarization direction**. Taking a basis in the transverse plane (the x–y basis for example, for light propagating in the z-direction), the two orthogonal components of $\boldsymbol{E}$ oscillate in phase with each other, reaching their maximum values at the same time. The polarization direction is determined by the ratio of the maximum amplitudes along the axes: $\boldsymbol{E}$ is along the line at angle $\theta = \tan^{-1} E_{0y}/E_{0x}$ from the x-axis. Light polarized at angles $\theta = 0$ and $\theta = \pi/2$ are, respectively, called horizontally (H) and vertically (V) polarized. Light polarized at $+\pi/4$ and $-\pi/4$ is said to be diagonally (D) and anti-diagonally (A) polarized (figure 2.5).

On the other hand, it is possible for the two components to be a quarter cycle out of phase, so that one component vanishes when the other is at a maximum; in this case, the electric field vector rotates as the light propagates. This is referred to as **elliptical polarization**. As the light propagates, the tip of the electric field vector traces out an ellipse, the **polarization ellipse**. If the elliptically polarized light has equal maximum amplitudes in both the x and y directions, then the light is **circularly polarized**. Depending on the direction the $\boldsymbol{E}$ vector is rotating, the light is said to be right-circular (R) or left-circular (L) polarized (figure 2.5).

A convenient means of describing the polarization state of light is through its Jones vector. Let E_x and E_y be magnitudes of the horizontal and vertical components of electric field. Writing the field in complex matrix form, we construct the two-component **Jones vector**,

$$\psi(z, t) = \begin{pmatrix} E_x \\ E_y\ e^{i\phi} \end{pmatrix} e^{-i(\omega t - kz)}, \qquad (2.50)$$

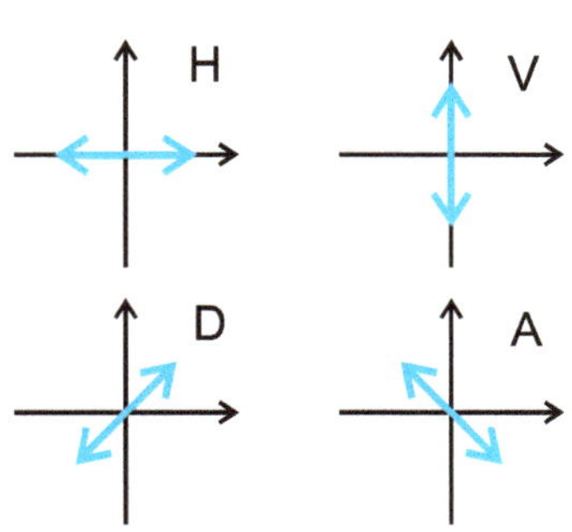

Figure 2.5. Linear polarization states of light: (a) horizontal, $|H\rangle$, (b) vertical, $|V\rangle$, and (c) diagonal, $|D\rangle$, (d) antidiagonal, $|A\rangle$. More generally, the light can be polarized at any angle θ from the x-axis, with angles θ and $\theta \pm \pi$ being equivalent to each other.

where ϕ is the phase difference between the components. Up to overall constants that we will ignore, the intensity of the light is given by the inner product of the Jones vector with itself,

$$I = \frac{1}{2}\psi^\dagger\psi = \frac{1}{2}\left(E_x^2 + E_y^2\right), \tag{2.51}$$

where the dagger represents the Hermitian conjugate (complex-conjugate transpose). The actions of optical devices such as lenses, waveplates, and polarizing filters can then be described as matrices acting on the incoming Jones vector.

We may then define a set of intensities for various polarization states:

$$I_H = \frac{1}{2}E_x^2 \tag{2.52}$$

$$I_V = \frac{1}{2}E_y^2 \tag{2.53}$$

$$I_D = \frac{1}{4}\left(E_x^2 + E_y^2\right) + \frac{1}{2}(E_xE_y)\cos\phi \tag{2.54}$$

$$I_A = \frac{1}{4}\left(E_x^2 + E_y^2\right) - \frac{1}{2}(E_xE_y)\cos\phi \tag{2.55}$$

$$I_R = \frac{1}{4}\left(E_x^2 + E_y^2\right) + \frac{1}{2}(E_xE_y)\sin\phi \tag{2.56}$$

$$I_L = \frac{1}{4}\left(E_x^2 + E_y^2\right) - \frac{1}{2}(E_xE_y)\sin\phi. \tag{2.57}$$

These represent the intensities for horizontal and vertical, diagonal and anti-diagonal, and right- and left-circular polarizations within the beam. The **Stokes parameters** are then defined to be

$$S_0 = I_H + I_V = \frac{1}{2}\left(E_x^2 + E_y^2\right) \tag{2.58}$$

$$S_1 = I_H - I_V = \frac{1}{2}\left(E_x^2 - E_y^2\right) \tag{2.59}$$

$$S_2 = I_D - I_A = E_xE_y\cos\phi \tag{2.60}$$

$$S_3 = I_R - I_L = E_xE_y\sin\phi. \tag{2.61}$$

These parameters have simple meaning. S_0 is just the total intensity, while the other three components measure various types of polarization: S_1, S_2, and S_3, respectively, measure the difference between vertical and horizontal polarizations, between the two diagonal polarizations, and between the two circular polarizations. For unpolarized light, $S_1 = S_2 = S_3 = 0$.

From the Stokes parameters one may form a three-dimensional column vector,

$$\mathcal{S} = \begin{pmatrix} S_1/S_0 \\ S_2/S_0 \\ S_3/S_0 \end{pmatrix}. \tag{2.62}$$

(Note that S_0, being the intensity, is non-negative.) The **degree of polarization** is defined to be

$$\mathcal{D} = |S| = \frac{\sqrt{S_1^2 + S_2^2 + S_3^2}}{S_0}. \tag{2.63}$$

Clearly, $\mathcal{D}$ vanishes for unpolarized light and reaches a maximum value of 1 for complete polarization.

A convenient pictorial representation of the polarization state of light is by means of the **Poincaré sphere**. Plotting the three-dimensional vector S, the unit vectors that represent the completely polarized states span the unit sphere, with partially polarized states in the interior of the sphere and completely unpolarized light at the center. With the choice of axes as shown in figure 2.6, horizontally polarized states lie at the North pole, vertically polarized at the South, and all other linearly polarized states on the great circle passing through the poles in the S_1-S_2 plane. The diagonal and anti-diagonal states are at the intersections of this circle with the S_2-axis. The circularly polarized states then lie on the equation in the S_2-S_3 plane. Two polarization states that are orthogonal to each other are always 180° apart, on opposite sides of the sphere (they are *antipodal*). In general, two linear polarization vectors at angle θ from each other in real space represent states that are separated by an angle of 2θ on the sphere.

An analog of the Poincaré sphere is also useful in the quantum mechanics of two-state systems, where it usually goes by the name of the **Bloch sphere.** In this context,

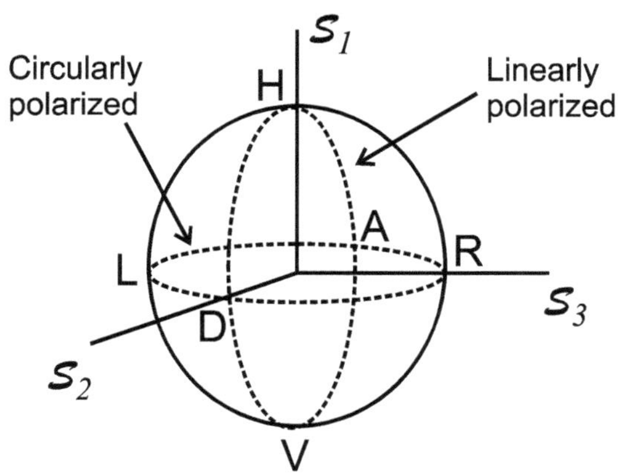

Figure 2.6. The Poincaré (or Bloch) sphere. Completely polarized states lie on the surface, with partially polarized states in the interior and completely unpolarized light at the center. All linearly polarized and circularly polarized states lie on the dashed circles.

pure quantum states lie on the surface of the sphere, while mixed states (those that are statistical ensembles of pure states) lie in the interior. The Bloch sphere enters into many applications in quantum information processing and in solid state physics [11–15].

As a polarization state is taken around a closed loop on the Poincaré sphere, it may gain an unexpected phase shift of purely geometric origin. This Pancharatnam phase [16–20] is discussed in section 9.1.

More detail on the physics of polarized light may be found in [2, 21, 22].

References

[1] Simon D S 2020 *A Guided Tour of Light Beams: From Lasers to Optical Knots* 2nd edn (Bristol: IOP Publishing)

[2] Saleh B E A and Teich M C 2019 *Fundamentals of Photonics* 3rd edn (Hoboken, NJ: Wiley)

[3] Aharonov Y and Bohm D 1959 *Phys. Rev.* **115** 485

[4] Peshkin M and Tonomura A 1989 *The Aharonov–Bohm Effect* (Berlin: Springer)

[5] Boyd R W 2008 *Nonlinear Optics* 3rd edn (San Diego, CA: Academic)

[6] Powers P E and Haus J W 2017 *Fundamentals of Nonlinear Optics* 2nd edn (Boca Raton, FL: CRC Press)

[7] Shen Y R 2003 *The Principles of Nonlinear Optics* (Hoboken, NJ: Wiley)

[8] Gerry C and Knight P 2004 *Introductory Quantum Optics* (Cambridge: Cambridge University Press)

[9] Ou Z Y 2007 *Multi-Photon Quantum Interference* (Berlin: Springer)

[10] Shih Y 2011 *An Introduction to Quantum Optics: Photon and Biphoton Physics* (Boca Raton, FL: CRC Press)

[11] Nielsen M A and Chuang I L 2014 *Quantum Computation and Quantum Information* (Cambridge: Cambridge University Press)

[12] Simon D S, Jaeger G and Sergienko A V 2014 *Int. J. Quant. Inf.* **12** 1430004

[13] Simon D S, Jaeger G and Sergienko A V 2017 *Quantum Metrology, Imaging, and Communication* (Berlin: Springer)

[14] Haroche S and Raimond J M 2006 *Exploring the Quantum: Atoms, Cavities, and Photons* (Oxford: Oxford University Press)

[15] Jaeger G 2007 *Quantum Information: An Overview* (Berlin: Springer)

[16] Pancharatnam S 1956 *Proc. Indian Acad. Sci.* A **44** 247

[17] Nitayananda R 1994 *Curr. Sci.* **67** 238

[18] Bhandari R 1994 *Curr. Sci.* **67** 224

[19] Ramachandran R and Ramaseshan S 1961 *Handbuch Der Physik* ed S Flugge vol. 12 (Berlin: Springer) 257 p

[20] Bhandari R 1997 *Phys. Rep.* **281** 1

[21] Damask J N 2005 *Polarization Optics in Telecommunications* (Berlin: Springer)

[22] Goldstein D H 2011 *Polarized Light* 3rd edn (Boca Raton, FL: CRC Press)

IOP Publishing

Topology in Optics (Second Edition)

Tying light in knots

David S Simon

Chapter 3

Characterizing spaces

In this chapter, some basic concepts of topology are introduced, focusing on aspects that are of immediate use in physics and optics. A few of the more formal aspects of topology are discussed briefly in appendix A; for more detailed discussions and for proofs, see topology textbooks such as [1–7].

3.1 Loops, holes, and winding numbers

Loosely speaking, two spaces are topologically equivalent to each other if they can be continuously deformed into each other. One of the most obvious ways to show that two spaces are *not* topologically equivalent is to show that they have different numbers of holes in them. For example, the single-holed torus and the double-holed torus are inequivalent: there is no way to continuously deform the former into the latter: to go from the single to the double torus it is necessary to tear the space to create the second hole, and tearing is a discontinuous transformation.

So how do you characterize the number of holes? A simple way is to look at the sets of closed loops that can be continuously deformed into each other. Consider a plane with a single hole punctured in it (left side of figure 3.1). The loops marked A and B can be continuously deformed into each other. In fact, they can both be continuously collapsed down to a single point. So we consider these loops to be equivalent to each other. However, loop C circles the hole. It can't be continuously deformed into either A or B (or to a single point), because it gets snagged on the hole. Similarly, a loop that circles the hole twice can't be deformed into A, B, *or* C. We therefore have an infinite set of equivalence classes of loops: the nth class consists of all the loops that circle the hole n times. The loops on this space are therefore characterized by a single integer, called the **winding number**, which will be defined in more detail in chapter 5. Note that the loop has an orientation given by the direction it rotates (clockwise or counterclockwise). The winding number has a sign determined by this orientation: we take $n > 0$ if the loop circulates around the hole counterclockwise, and $n < 0$ for clockwise.

doi:10.1088/978-0-7503-3471-6ch3

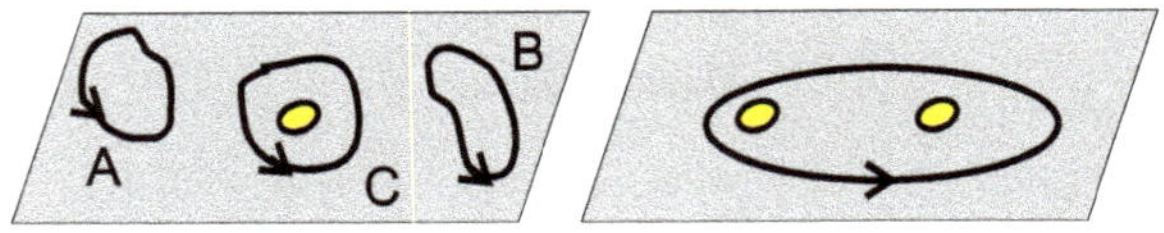

Figure 3.1. On the left, the plane with a single puncture in it contains loops that cannot be continuously deformed into each other without getting caught on the hole. So each loop can be characterized by an integer counting the number of times n the loop is circled. For the loops shown, A and B are equivalent to each other, with $n = 0$. C is not equivalent to the other two loops, since it has $n = 1$. For the plane on the right, with two punctures, loops are characterized by a *pair* of integers (n_1, n_2) counting the number of times each of the two holes is enclosed; the loop shown has $n_1 = n_2 = 1$.

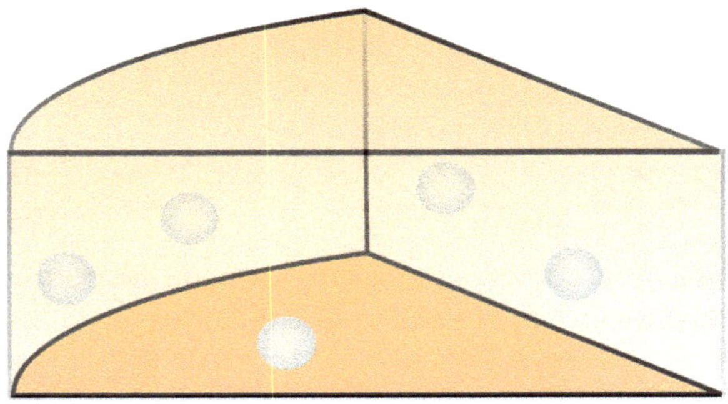

Figure 3.2. The air bubbles in a piece of Swiss cheese are a different type of hole than the puncture in the plane. Whereas a loop (which is deformable to a circle) can slide around any of the bubbles and be deformed into any other loop, a two-dimensional sphere that surrounds a bubble cannot be deformed into a sphere that doesn't. So the space is characterized by equivalence classes of two-dimensional spheres, rather than one-dimensional loops.

But now consider a second plane, with two holes punctured in it (right side of figure 3.1). Here, characterizing equivalence classes of loops now requires specifying two integers, (n_1, n_2), where n_1 counts how many times the loop circles the left-hand hole and n_2 counts the number of windings around the right-hand hole. A plane with three holes would require specifying three integer winding numbers, and so on.

This is still not the end of the story, however. A space may have holes of different types. Holes in the plane are not the same as holes in a piece of Swiss cheese (figure 3.2). A loop like those of figure 3.1 can slip around the holes in the cheese. So loops are incapable of detecting these 'higher-dimensional' holes. However, instead of loops (which are topologically equivalent to a circle or one-dimensional sphere, S^1) we can use two-dimensional spheres, S^2 and look at equivalence classes of spheres that can be continuously deformed into each other without crossing holes. So equivalence classes of spheres that wind around all of the holes the same number of times can be used to characterize the space. These integer spherical winding numbers will be called Chern numbers (chapter 5).

Continuing in this way, we can characterize the hole structure of a space by evaluating equivalence classes of spheres of different dimensions. A method for testing whether or not two spaces are topologically equivalent is then apparent: compare the list of such equivalence classes on spheres for the two spaces. If the lists are not equal, then the spaces are distinct and cannot be smoothly deformed into each other. Within a given space, each sphere is then associated with a sequence of integers corresponding to its windings about all of the holes, and spheres with differing sets of integer values are topologically inequivalent.

The idea of treating spaces with the same hole structure as equivalent is formalized by the idea of **homotopy classes**, which will be defined in the next section. Winding numbers and other topological invariants will be discussed in more detail in chapter 5.

3.2 Homotopy classes

Homotopy classes are a means of classifying the structure of a topological space by equivalence classes of circles (one-dimensional loops), or more generally, of spheres of different dimensions. The idea of classifying surfaces by means of loops apparently goes back to Camille Jordan in the 1860s. The idea of imposing a group structure on the set of homotopy classes originated with the work of Poincaré in the 1890s.

Let X be a topological space (see the appendix), and let I denote the unit interval, $I = [0, 1]$. A **path** α in X with endpoints x_0 and x_1 is a continuous map α: $I \to X$ (in other words the map has image $\alpha(t)$ which forms a curve in X, for $0 \leqslant t \leqslant 1$), such that $\alpha(0) = x_0$ and $\alpha(1) = x_1$ (figure 3.3(a)). A **loop** in X is a path whose ends are identified, $x_0 = x_1$ (figure 3.3(b)). The loop is said to be **based** at x_0.

Given two loops α and β based at the same point, one can define a product path $\alpha * \beta$ as the path that follows one loop until it returns to the base point, then follows the other loop (figure 3.3(c)):

$$\alpha * \beta(t) = \begin{cases} \alpha(2t) & \text{for} \quad 0 \leqslant t \leqslant \dfrac{1}{2} \\ \beta(2t - 1) & \text{for} \quad \dfrac{1}{2} < t \leqslant 1. \end{cases} \tag{3.1}$$

The constant loop is simply the loop that stays fixed at x_0 for all t: $\alpha(t) = x_0$ for $0 \leqslant t \leqslant 1$, and the inverse of the loop is obtained by running the parameter t in the opposite direction: $\alpha^{-1}(t) = \alpha(1 - t)$. With these definitions, it can easily be shown that the set of loops based at a given point form a mathematical group, with the constant loop playing the role of the group identity element.

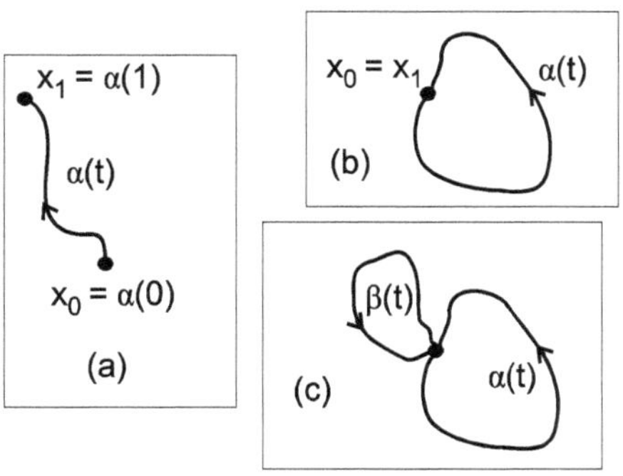

Figure 3.3. (a) A path going from x_0 to x_1. (b) A closed loop obtained by identifying the two endpoints of a path. (c) The product of two loops.

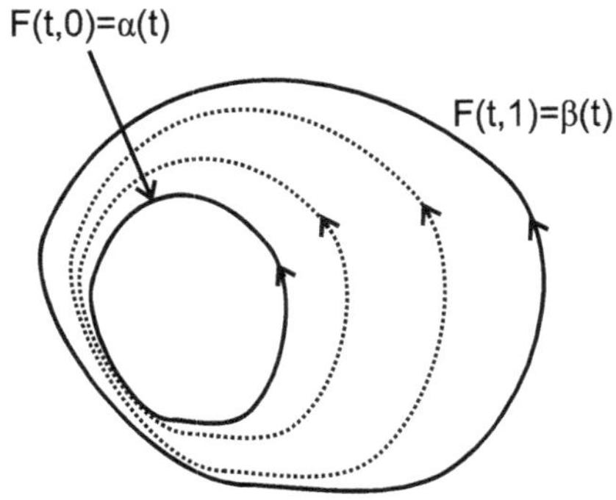

Figure 3.4. Homotopy of two loops $\alpha(t)$ and $\beta(t)$. The innermost loop is $F(t, 0) = \alpha(t)$, the outermost loop is $F(t, 1) = \beta(t)$. The dotted loops are representative examples of $F(t, s)$ for two other values of s $(0 < s < 1)$. As s increases, $\alpha(t)$ gradually evolves into $\beta(t)$.

But a more interesting and useful group may be obtained by adding a second parameter. Given two loops $\alpha(t)$ and $\beta(t)$ in X, based at the same point, we define a two-parameter continuous map F: $I \times I \to X$, which provides a continuous deformation of α into β:

$$F(t, 0) = \alpha(t), \qquad F(t, 1) = \beta(t), \qquad F(0, s) = F(1, s) = x_0 \tag{3.2}$$

(see figure 3.4). The first parameter t carries us along the loop, while varying the second parameter continuously deforms one loop into the other. Such a deformation is called a **homotopy**. The idea of homotopy can be generalized in an obvious manner from loops to arbitrary continuous maps.

If two loops α and β at x_0 are homotopic to each other, we write $\alpha \sim \beta$. Homotopy is an equivalence relation, so we define the **homotopy classes** $[\alpha]_{x_0}$ of loop α to be the equivalence class of loops homotopic to α. In other words $[\alpha]_{x_0}$ is the set of loops at x_0 continuously deformable into α.

Given two topological spaces, X and Y, they are said to be of the same **homotopy type** if there exist continuous maps f: $X \to Y$ and g: $Y \to X$ such that

$$f \circ g \sim \mathcal{I}_Y \quad \text{and} \quad g \circ f \sim \mathcal{I}_X, \tag{3.3}$$

where $\sim$ denotes equivalence under homotopy and $\mathcal{I}_{X,Y}$ denote the identity maps on spaces X and Y. In other words, f and g are inverses up to homotopy.

A topological space X is called **arc-wise connected** if, given any two points $x_0, x_1 \in X$, there is a path such that x_0 and x_1 are its endpoints. On an arc-wise connected space, the set of homotopy classes at each base point is isomorphic to the homotopy classes at any other base point. In this case, all base points are equivalent, and so there is no need to specify which base point is used. Henceforth, we only consider arc-wise connected spaces and will usually omit mention of the base point.

Given the product of loops defined above, the set of homotopy classes inherits a natural product, which will also be denoted *:

$$[\alpha] * [\beta] = [\alpha * \beta]. \tag{3.4}$$

Inverses, associativity and the existence of an identity element then follow:

$$[\alpha]^{-1} = [\alpha^{-1}], \tag{3.5}$$

$$([\alpha] * [\beta]) * [\gamma] = [\alpha] * ([\beta] * [\gamma]), \tag{3.6}$$

$$[\alpha] * [c] = [c] * [\alpha] \ = \alpha, \tag{3.7}$$

where the equivalence class $[c]$ of the constant loop c serves as the identity element. The homotopy classes therefore form a group, called the **fundamental group** or the **first homotopy group** of X, denoted by $\pi_1(X, x_0)$ (including base point) or simply as $\pi_1(X)$ (if base point is suppressed). It can be shown that two spaces of the same homotopy type have the same fundamental group.

Consider some simple examples:

(i) **Euclidean spaces**, $\mathbb{R}^n$. All loops are deformable to each other, so there is a single homotopy class. The homotopy group has a single entry, and since every group must contain the identity element $\mathcal{I}$, the group in this case consists of *just* the identity: $\pi_1(\mathbb{R}^n) = \{\mathcal{I}\}$.

(ii) The **n-dimensional sphere**: S^n. For $n > 1$, all loops on the sphere can be smoothly deformed into each other simply by sliding them around on the sphere's surface, so that once again $\pi_1(S^n) = \{\mathcal{I}\}$ for $n > 1$. However, for $n = 1$, the one-dimensional sphere is a circle; closed loops on the circle are distinguished from each other by a single integer, the number of times they wind around the circle. So the homotopy group is simply the group of integers: $\pi_1(S^1) = \mathbb{Z}$.

(iii) **Tori:** An n-dimensional torus T^n is formed from the product of n circles, so there are n integers counting the windings about each hole. Therefore, $\pi_1(T^n) = \mathbb{Z}^n$.

Evaluating the fundamental group allows us to detect holes such as those in a punctured plane or the hole in a donut. As a more physical example, the interior of the solenoid in the Aharonov–Bohm (AB) effect serves as a hole in the charged particle's configuration space, so that the AB effect can be viewed as a consequence of the nontrivial first homotopy group. But as mentioned earlier, there are other types of holes that cannot be detected by looking at deformations of circles. To study these holes, we must move from circles (one-dimensional spheres) to spheres of higher dimension; this leads us to define the higher homotopy groups.

Let the symbol ∂ denote the boundary operator; ∂M is the set of points on the boundary of M. Consider a unit cube, the set of points

$$I_n = \{s_1, s_2, \dots, s_n\}, \quad \text{with} \quad 0 \leqslant s_i \leqslant 1. \tag{3.8}$$

The boundary of I_n is the surface of the cube. In one dimension, the unit interval $I = I_1$ has boundary given by the pair of endpoints, $\partial I_1 = \{0, 1\}$; this interval can be converted into a circle by identifying the endpoints with each other, or in other words, gluing the ends of the interval together (figure 3.5). Similarly, for $n > 1$ we can collapse the boundary of I_n to a single point (identify all points on the surface

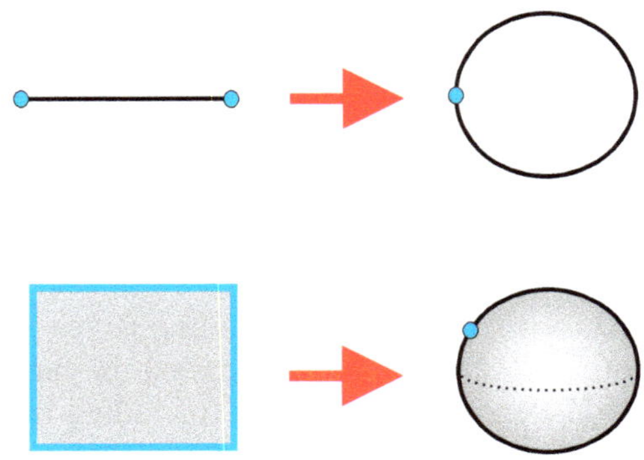

Figure 3.5. Forming n-spheres by identifying the boundaries of n-cubes. For $n = 1$, the endpoints of the unit interval or 1-cube (the blue dots) are identified with each other to form a circle. For $n = 2$, the boundary of the square is collapsed to a point to form a 2-sphere.

with each other), to convert the n-cube into something isomorphic to an n-sphere, S^n. This identification can be formally written as $S^n = I_n/\partial I_n$. Now that ∂I_n is collapsed to a point, x_0, we can use it as a base point for n-loops. The ***n*-loop** α at x_0 is a continuous map of the n-dimensional cube to topological space X, leaving the boundary fixed:

$$\alpha\colon I_n \to X, \qquad \text{such that} \quad \alpha\colon \partial I_n \to x_0. \tag{3.9}$$

Two such loops are then **homotopic** $\alpha \sim \beta$ if there exists a **homotopy** $F\colon I_n \times I \to X$ such that:

$$F(s_1, \dots, s_n, 0) = \alpha(s_1, \dots, s_n) \tag{3.10}$$

$$F(s_1, \dots, s_n, 1) = \beta(s_1, \dots s_n) \tag{3.11}$$

$$F(s_1, \dots, s_n, t) = x_0 \quad \text{for} \quad (s_1, \dots, s_n) \in \partial I_n. \tag{3.12}$$

The homotopy relation $\alpha \sim \beta$ is again an equivalence relation, so that we may define the corresponding homotopy equivalence classes $[\alpha]$. The product of n-loops, $\alpha * \beta$ is

$$\alpha * \beta(s_1, \dots, s_n) = \begin{cases} \alpha(2s_1, s_2, \dots, s_n) & \text{for} \quad 0 \leqslant s_1 \leqslant \dfrac{1}{2} \\ \beta(2s_1 - 1, s_2, \dots, s_n) & \text{for} \quad \dfrac{1}{2} < s_1 \leqslant 1. \end{cases} \tag{3.13}$$

The ***n*th homotopy group** at x_0 for $n \geqslant 2$ is then defined in direct analogy to the fundamental group: $\pi_n(X, x_0)$ is the group of equivalence classes of continuous maps $S^n \to X$, where we are identifying $I_n/\partial I_n$ with the n-sphere. π_n quantifies the set of topologically inequivalent n-spheres in the topological space that cannot be deformed into each other without being obstructed by holes.

We have now defined homotopy groups $\pi_n(X)$ for $n \geqslant 1$. We can carry out the analogous construction in zero dimensions as well to form a zeroth homotopy 'group': the zero-dimensional interval is simply a point, $I_0 = x_0$, with the boundary being the empty set: $\partial I_0 = \{\varnothing\}$. The zero-sphere $S^0 \sim I_0/\partial I_0$ is then the point x_0, and all loops are just constant maps. Two such loops at points x and y in X will be homotopic, $\alpha \sim \beta$, if and only if x and y can be smoothly deformed into each other, i.e. if they are the endpoints of some continuous curve. A space may be composed of multiple components that are disconnected from each other, such as several concentric spheres nested inside each other. The set of equivalence classes, $\pi_0(X)$ is then just the set of connected components. A space is simply connected if $\pi_0(X)$ is the trivial group containing a single element, while a multiply connected space has $\pi_0(X)$ isomorphic to a finite set of integers that label the connected components. Note however, that $\pi_0(X)$ is **not** a group, unlike the $\pi_n(X)$ with $n > 0$.

A stronger notion of topological equivalence than homotopy type is **homeomorphism**, which means essentially that two spaces can be continuously deformed into each other. (See the appendix for the precise definition.) As mentioned above, spaces with the same fundamental group are of the same homotopy type. However, since there are types of holes that cannot be distinguished by the fundamental group, being of the same homotopy type is a weaker condition than being homeomorphic. Dimension is largely invisible to the fundamental group. For example, points and circles are homotopy equivalent to solid balls and Möbius bands, respectively, due to the ability to continuously contract one to the other; however these spaces are not homeomorphic (note that the contraction is not uniquely invertible). Including the higher homotopy classes is one way to help distinguish between spaces that are of the same homotopy type but which are not homeomorphic.

A number of commonly used homotopy groups for spheres, tori, and Lie groups are tabulated in the appendix.

References

[1] Fulton W 1995 *Algebraic Topology: A First Course* (Berlin: Springer)

[2] Hatcher A 2002 *Algebraic Topology* (Cambridge: Cambridge University Press)

[3] Greenberg M and Harper J 2018 *Algebraic Topology: A First Course* (Boca Raton, FL: CRC Press)

[4] Munkres J R 2000 *Topology* (Upper Saddle River, NJ: Prentice-Hall)

[5] Hirsch M W 1997 *Differential Topology* (Berlin: Springer)

[6] Guillemin V and Pollock A 1974 *Differential Topology* (Englewood Cliffs, NJ: Prentice-Hall)

[7] Basener W F 2006 *Topology and Its Applications* (Hoboken, NJ: Wiley)

IOP Publishing

Topology in Optics (Second Edition)
Tying light in knots
David S Simon

Chapter 4

Fiber bundles, curvature, and holonomy

In this chapter, we give a brief introduction to several topics in geometry and topology. The principal mathematical objects introduced in this chapter, curvature, holonomy, and fiber bundles, arise in many physical applications.

Curvature is a local, geometric quantity; it changes under continuous deformations of the surface, and so is not a topological invariant. Despite this, it is relevant to topology, since topological invariants can be constructed from it. These invariants need to be globally defined quantities, so they usually involve integrals or averages of the curvature over the full manifold. Invariance of the topology places global constraints on the curvature: increases of curvature at one point of the manifold must be compensated for by decreases at other points. In this way, local geometry and global topology tend to become linked.

The most famous example of a topological invariant related to curvature is the Euler–Poincaré characteristic, $\chi(M)$. This and other invariant quantities will be discussed in the next chapter, which will in turn bring the discussion back to the homotopy groups of chapter 3. In the current chapter, some of the geometric prerequisites are covered.

Many excellent textbooks and review articles exist on the topics covered in this chapter. Just to list a few, introductions to differential geometry include [1–3], while introductions to differential topology may be found in [4, 5]. Fiber bundles are treated in detail in [6]. More physics-oriented treatments of many of these topics may be found in [7–10]. Excellent introductions to the geometric approach to gauge theories like electromagnetism include [11, 12].

4.1 Manifolds

Roughly speaking, a manifold is a space which is sufficiently well-behaved that you can do calculus on it. More precisely, M is a **differentiable manifold** of dimension n if

doi:10.1088/978-0-7503-3471-6ch4

the following conditions hold (see appendix A.1 for the definitions of topological spaces, open sets, and homeomorphisms):

1. M is a topological space.
2. M is covered by a collection of open set U_i, $i = 1, 2, \ldots$, such that the union of all the sets equals M.
3. There is a collection of mappings (homeomorphisms) f_i that map the open sets continuously and invertibly to open sets of the cartesian space $\mathbb{R}^n$, $f_i: U_i \to \mathbb{R}$.
4. On the overlap of two open sets U_i and U_j, the maps $\psi_{ij} = f_i^{-1} \circ f_j$ are differentiable.

The open sets in this definition can provide **coordinate patches**, regions on which a single coordinate system can be provided, with the f_i mapping the coordinates on U_i to coordinates on a region of $\mathbb{R}^n$, $f_i: U_i \to \mathbb{R}^n$. The definition allows for multiple overlapping coordinate patches (multiple open set) because many manifolds can't be consistently covered with a single coordinate system. For example the two-dimensional sphere S^2 requires at least two overlapping coordinate patches; any attempt to cover the sphere with a single coordinate system will lead to the coordinate axes becoming ill-defined at some point (see figure 4.1). In this case, coordinate systems can only be defined locally on each U_i and then patched together. The patching together is done by the **transition functions** $\psi_{ij} = f_i^{-1} \circ f_j$, which allow for smooth changes of coordinate system from the coordinates on

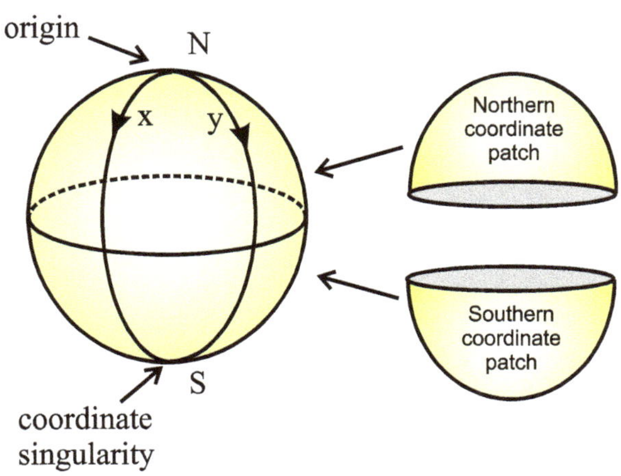

Figure 4.1. A single coordinate patch cannot cover the sphere. For example, define a coordinate system $\{x, y\}$ with origin at the North pole. The coordinate axes start out orthogonal, and as one moves away from the origin, the coordinate system remains well-defined until the South pole is reached, where the two coordinate axes collide. Any other choice of coordinates will similarly have at least one such singular point; for example, in a latitude–longitude system the longitude becomes undefined at both poles. A solution is to define two hemispherical coordinate patches, one with origin at the North pole, the other with origin at the South pole. The overlap of the patches is a small band at the equator, on which transition functions ψ_{ij} translate between the two coordinate systems.

U_j to those of U_i on the regions where the sets overlap. Common examples of manifolds include n-dimensional spheres S^n, tori T^n, and Euclidean spaces $\mathbb{R}^n$.

Given an m-dimensional differential manifold M, let T_pM denote the **tangent space** at point p in M: T_pM is the set of vectors tangent to M at point p. The manifold itself is called a **Riemannian manifold** if it has associated to it a **Riemannian metric** g, a continuous tensor of type (2,0) with the following properties:

(a) g is symmetric: $g(V, W) = g(W, V)$ for all vectors $V, W \in T_pM$.
(b) g is nondegenerate: $g(V, W) = 0$ for all $V \in T_pM$ implies that $W = 0$.
(c) If the metric is expanded in local coordinates $\{x^\mu\}$ as $g = g_{\mu\nu} dx^\mu \otimes dx^\nu$, the matrix $g_{\mu\nu}$ has positive eigenvalues.

In (c), dx^μ is the one-form defined in the next section, and $\otimes$ is the (symmetric) tensor product. We have also used the **Einstein summation convention**, in which an index that is repeated (once up and once down) is summed over.

The metric provides an inner product, a distance measure, and as will be seen below a way of raising and lowering indices.

If condition (c) is relaxed to allow negative eigenvalues as well, the resulting tensor is called a **pseudometric** (usually simply called a metric in the physics literature), and the manifold is called **semi-Riemannian**. Pseudometrics commonly occur in relativistic physics.

It will be helpful to look at derivatives in a different way, rather than as arrows in space. Consider a function f defined on some flat space with coordinates $x_1, x_2, \ldots$ and basis vectors $\hat{e}_1, \hat{e}_2, \ldots$, etc. Imagine that we wish to evaluate the function as we displace our measurement point from some original location $\boldsymbol{r}_0 = \sum_i r_0^i \hat{e}_i$ to some new, nearby point $\boldsymbol{r}_0 + \delta\boldsymbol{r}$: $f(\boldsymbol{r}_0) \to f(\boldsymbol{r}_0 + \delta\boldsymbol{r})$. This can be done using a Taylor series, expanding the function about $\boldsymbol{r}_0$. This series can be recast into a different form by thinking of the derivatives as operators that can be moved around and exponentiated like variables:

$$f(\boldsymbol{r}_0 + \delta\boldsymbol{r}) = \sum_{n=0}^{\infty} \frac{(\delta r^i)^n}{n!} \left(\frac{\partial}{\partial x^i}\right)^n f(\boldsymbol{r}) \Big|_{\boldsymbol{r}_0} \tag{4.1}$$

$$= e^{\delta r^i (\partial/\partial x^i)} f(\boldsymbol{r}_0). \tag{4.2}$$

The operator $\exp(\delta r^i(\partial/\partial x^i))$ is a **translation operator**, shifting the point at which the function is evaluated through a displacement $\delta\boldsymbol{r}$. The derivatives in the exponential are said to be the **generators** of translations. The vectors $\delta r^i \hat{e}_i$ are in one-to-one correspondence with the linear differential operators $\delta r^i(\partial/\partial x^i)$, so it is common in geometry to equate the derivative operator with the vector. In other words, a vector v with components v^i is written as

$$v = \sum_i v^i \partial_i, \tag{4.3}$$

where ∂_i is short-hand for $\partial/\partial x^i$. The partial derivatives along the coordinate axes serve as basis vectors.

Given a curve $\mathbf{r}(t)$ with parameter t, the tangent vector along the curve is then written by means of the chain rule as

$$v = \frac{d\mathbf{r}}{dt} = \sum_i \frac{\partial r^i}{\partial t}\partial_i, \tag{4.4}$$

which is a vector with components $\partial r^i/\partial t$.

4.2 Vectors and forms

The space of linear functionals on T_pM (the set of linear mappings from T_pM to the real numbers, $\mathbb{R}$), is called the **cotangent space** or **space of dual vectors** T_p^*M at p. Given a local set of coordinates $\{x^\mu\}$ on M, the partial derivative operators $\{\partial/\partial x_\mu\}$ and the differentials $\{dx^\mu\}$, respectively, define bases of T_pM and T_p^*M, often called **coordinate bases**. Any basis not obtained in this way from a set of coordinates on M is called a **noncoordinate basis**.

On flat spaces like $\mathbb{R}^n$, the spaces T_pM and T_pM^* are often treated as interchangeable and no distinction is made between them. More generally, they are isomorphic to each and the metric can be used to define mappings between them:

$$\flat\colon T_pM \to T_p^*M$$
$$\sharp\colon T_p^*M \to T_pM.$$

Under these mappings, a vector with components V^α defines a one-form $V^\flat = V_\mu^\flat dx^\mu$ with components

$$V_\mu^\flat = g_{\mu\nu}V^\nu. \tag{4.5}$$

Similarly, a one-form ω with components ω_α defines a vector $\omega_\sharp = \omega_\sharp^\mu \partial_\mu$ with components

$$\omega_\sharp^\mu = g^{\mu\nu}\omega_\nu. \tag{4.6}$$

Most often in physics, the $\sharp$ and $\flat$ symbols are omitted, and the type of object is simply denoted by the positions of the indices, up or down. In this case, we would then simply write

$$V_\mu = g_{\mu\nu}V^\nu \tag{4.7}$$

and

$$\omega^\mu = g^{\mu\nu}\omega_\nu, \tag{4.8}$$

with the metric being used to raise and lower indices.

One-form and vectors map each other to scalars via **contraction**; for a one-form ω with components ω_μ and a vector V with components V^μ, the contraction is

$$\omega(V) = V(\omega) = V^\mu\omega_\mu = g_{\mu\nu}V^\mu\omega^\nu = g^{\mu\nu}V_\mu\omega_\nu. \tag{4.9}$$

The **tangent bundle** TM of M is the union of all the tangent spaces, $TM = \bigcup_{p\in M} T_pM$, and the **cotangent bundle** is $T^*M = \bigcup_{p\in M} T_p^*M$. A **vector field** is a continuous function on M with values in TM, while a **differential one-form** or **covector field** is a continuous function with values in T^*M. A (coordinate or non-coordinate) basis for TM is of the form $\{x^\mu, \hat{e}_\mu(x)\}$. Here $\{\hat{e}_\mu(x)\}$ is a basis of T_xM for each x, and is called a **moving frame** or vielbein; in the specific cases of two or four dimensions, it is called a zweibein or **vierbein**, respectively. For a coordinate basis $\hat{e}_\mu = \partial/\partial_x\mu$, while for any non-coordinate basis $\{\hat{e}_\mu(x)\}$, there is a matrix $\{e_a^{\ \mu}\}$ such that

$$\hat{e}_a(x) = e_a^{\ \mu}\frac{\partial}{\partial x^\mu}. \tag{4.10}$$

Suppose M is a differentiable manifold of dimension m, and consider the differentials dx^i, $i = 1, 2, \ldots, m$. The dx^i form a basis for the set of **one-form**. Define the **wedge product** or **exterior product**, an antisymmetric product on differentials:

$$dx^i \wedge dx^j = dx^i \otimes dx^j - dx^j \otimes dx^i. \tag{4.11}$$

The wedge product gives a two-form, which plays the role of an oriented two dimensional area (figure 4.2). Using the wedge product again gives a three-form and so on. The set of wedge products of r distinct differentials forms a basis for the space of r-forms. A **differential r-form** is a completely antisymmetric covariant tensor field of rank r. In other words, a tensor field $\omega(x)$ is a differential r-form if it can be written as $\omega(x) = (1/r!)\omega_{i_1,i_2,\ldots,i_r}(x)dx^{i_1} \wedge dx^{i_2} \wedge \cdots \wedge dx^{i_r}$, where the components $\omega_{i_1,i_2,\ldots,i_r}$ are antisymmetric under interchange of any two indices. Ordinary functions can be thought of as 0-forms. $\Lambda^r(M)$ is the set of differential r-forms, and the **exterior algebra** is the collection of all differential forms of all orders,

$$\Lambda(M) = \Lambda^0(M) \oplus \Lambda^1(M) \oplus \cdots \oplus \Lambda^m(M), \tag{4.12}$$

equipped with the operations of addition and exterior product.

On an n-dimensional manifold, the oriented **volume form** created by taking the product of all n one-forms, $dx^1 \wedge dx^2 \wedge \cdots \wedge dx^n$ allows integration of functions on

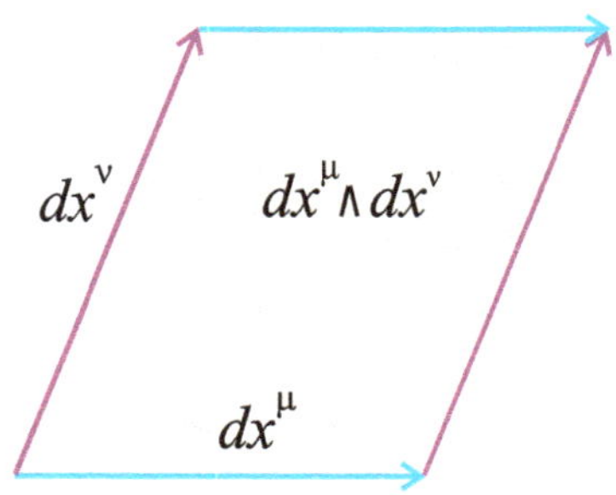

Figure 4.2. The antisymmetric wedge product of two differentials (one-forms) is an oriented two-dimensional area (a two-form). The orientation of the example shown can be thought of as a vector pointing out of the page, like a cross product. Reversing the order introduces a minus sign, $dx^\mu \wedge dx^\nu = -dx^\nu \wedge dx^\mu$, which reverses the orientation. Continuing to take the wedge product gives oriented volumes of higher dimensions.

open sets $U \subset M$: $\int_U f(x)\, dx^1 \wedge dx^2 \wedge \cdots \wedge dx^n$ is well-defined. Similarly, n-forms $\omega = \omega_{i_1,\dots i_n} dx_{i_1} \wedge \cdots \wedge dx_{i_n}$ have a volume-form already built into them, allowing them also to be integrated over U: $\int_U \omega$. Given an open cover of sets on M, the integrals can then be extended from the individual open sets to the full manifold.

The **exterior derivative**, $d\omega$, of an r-form ω, is the $(r + 1)$-form $d\omega = 1/r!(\partial_j \omega_{i_1, i_2, \dots, i_r})\, dx^j \wedge dx^{i_1} \wedge \cdots \wedge dx^{i_r}$. The exterior derivative generalizes the usual notions of divergence, curl, and gradient to arbitrary manifolds. In three-dimensional Euclidean space, vectors and one-forms can be treated as interchangeable since the metric is simply the identity matrix, but in general, we must distinguish between them. We find that d acts on 0-forms (functions) to produce the **gradient**:

$$\nabla f = (df)^{\sharp} \qquad \text{or} \qquad df = (\nabla f)^{\flat}.$$

On one-forms in three dimensions, d produces the **curl**:

$$\nabla \times V = [* \, d(V^{\flat})]_{\sharp}, \tag{4.13}$$

where the **Hodge star operator**, $*$, replaces dx^{μ} with the $(n - 1)$-form created by starting from the volume form and dropping dx^{μ} (see [7, 8, 10] for a more precise definition). On two-forms, the action of d is to give back the **divergence**:

$$\nabla \cdot V = * \, (d\omega), \tag{4.14}$$

where $\omega = *(V^{\flat})$ is a two-form. Due to the antisymmetry of the wedge product, d is a **nilpotent** operator, meaning that $d^2 = 0$; in three dimensions, this expresses the fact that the divergence of a curl vanishes.

If f is a one-dimensional function, f: $S \to \mathbb{R}$, then the differential df acting on a vector v gives the **directional derivative** of function f along the direction of v. It is related to the gradient by

$$df(v) = \langle \nabla f, v \rangle = v^i \partial_i f = v(f), \tag{4.15}$$

where the bracket denotes inner product.

A useful result is the generalized **Stokes' theorem**: given a p-form ω to be integrated over a $p + 1$-dimensional region U, we have

$$\int_U d\omega = \int_{\partial U} \omega, \tag{4.16}$$

where ∂U is the p-dimensional boundary of U.

4.3 Curvature

We wish to have a means of describing how a manifold curves. To do this requires defining a few preliminary notions. Given an n-dimensional manifold M, any point can be specified by a set of n numbers, the coordinates relative to the appropriate local coordinate system. Any curve on M can then be described by a set of functions $\{f_1(t), f_2(t), \dots, f_n(t)\}$, where t is a parameter along the curve and the functions give the coordinates of the point on the curve at parameter value t. (More precisely, what

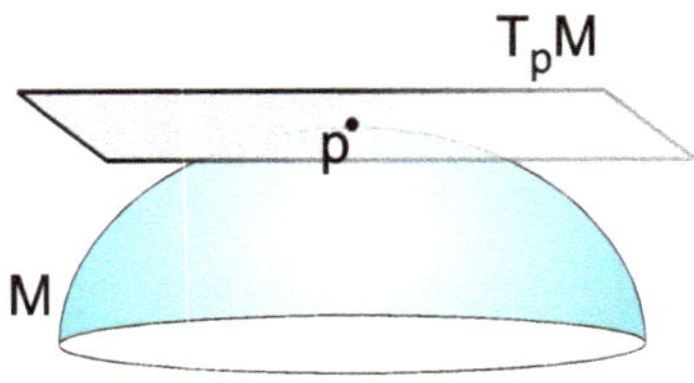

Figure 4.3. The tangent plane T_pM of manifold M at point p is spanned by all vectors tangent to M at p. It forms a linear or flat approximation to the manifold in the vicinity of p.

we are calling $f_i(t)$ should be written as $f_i(\alpha(t))$, where f_i is the coordinate mapping defined in the section 4.1 and $\alpha(t)$ defines the curve.) These n coordinate functions can of course be combined into a single vector-valued function $\boldsymbol{f}(t)$. The 'velocity' vector or **tangent vector** to the curve at t is the vector $\boldsymbol{v}(t) = \boldsymbol{f}'(t)$ whose components are $\{df_1/dt, df_2/dt, \ldots, df_n/dt\}$. At a given point p, the **tangent space** T_pM is the n-dimensional plane spanned by the tangent vectors of all possible curves in M passing through p. The tangent plane provides a linear (or flat) approximation to M in a neighborhood of p; it can be thought of as a piece of $\mathbb{R}^n$ sewn onto M at p (figure 4.3). The curvature at p can then be viewed as a measure of how fast M pulls away from the flat tangent space as one moves away from p. Before generalizing, first consider the simplest case, in which the manifold itself is a one-dimensional curve.

4.3.1 One-dimension: curves

Consider a one-dimensional curve in n-dimensional space, $\mathbb{R}^n$. This curve can be viewed as a map $\boldsymbol{f}: \mathbb{R} \to \mathbb{R}^n$, defined by the coordinate functions, $\{f_1(t), f_2(t), \ldots, f_n(t)\}$. The velocity vector $\boldsymbol{v}(t) = \boldsymbol{f}'(t)$ is obviously always tangent to the curve, pointing in the direction of increasing t. The magnitude, $v(t) = |\boldsymbol{v}(t)|$ is the speed of the curve in the given parameterization. The curve is said to have a **critical point** at t if $v(t) = 0$; otherwise it is **regular** there.

The **arclength** traversed during interval $(0, t)$ is

$$s(t) = \int_{t_0}^{t} |\boldsymbol{f}'(t')|\, dt' = \int_{0}^{t} \left|\frac{\partial f}{dt'}\right| dt'. \tag{4.17}$$

The derivative of arclength with respect to the parameter equals the speed:

$$\frac{ds}{dt} = |\boldsymbol{f}'| = v. \tag{4.18}$$

Often, curves are parameterized by arclength, $t = s$, so that they are of unit speed, $v(t) = 1$ for all t. Henceforth, we assume such arclength parameterization unless stated otherwise.

The **curvature** k of the curve is the magnitude of the acceleration:

$$k(s) = |\boldsymbol{f}''(s)| = |\boldsymbol{v}'(s)|, \tag{4.19}$$

or in other words, the rate at which the tangent vector's direction changes as one moves along the curve. The **radius of curvature** at any point is given by

$$R(s) = \frac{1}{k(s)}. \tag{4.20}$$

Also notice that for arc-length-paramaterized curves,

$$f' \cdot f'' = \frac{1}{2}\frac{d}{ds}(f' \cdot f') = \frac{d}{ds}|f'|^2 = \frac{d}{ds}(1) = 0, \tag{4.21}$$

so f'' and $v = f'$ are always perpendicular to each other. Therefore, if we define

$$f'' = k\boldsymbol{n}, \tag{4.22}$$

then $\boldsymbol{n}$ is a unit vector perpendicular to the tangent vector, $v \cdot \boldsymbol{n} = 0$, with $\boldsymbol{n} \cdot \boldsymbol{n} = 1$. $\boldsymbol{n}$ is the **normal vector** of the curve.

In three dimensions, one may define an additional unit vector, $\boldsymbol{b}$, called the **binormal vector**, perpendicular to both the tangent and normal vectors. We can take that vector to be

$$\boldsymbol{b}(s) = v(s) \times \boldsymbol{n}(s). \tag{4.23}$$

So now we have a set of three orthonormal basis vectors $\{v, \boldsymbol{n}, \boldsymbol{b}\}$ (figure 4.4), that bend and follow the curve. If the curve exists in a three-dimensional space, it can be completely characterized by these vectors. Together the three unit vectors are called the **Frenet trihedron** or **Frenet–Seret frame**. They are defined so that

$$\boldsymbol{n} = \boldsymbol{b} \times v \tag{4.24}$$

$$v = \boldsymbol{n} \times \boldsymbol{b} \tag{4.25}$$

$$\boldsymbol{b} = v \times \boldsymbol{n}. \tag{4.26}$$

We can also look at the derivatives with respect to s of the three unit vectors. The derivative of v was already given above:

$$v' = k\boldsymbol{n}. \tag{4.27}$$

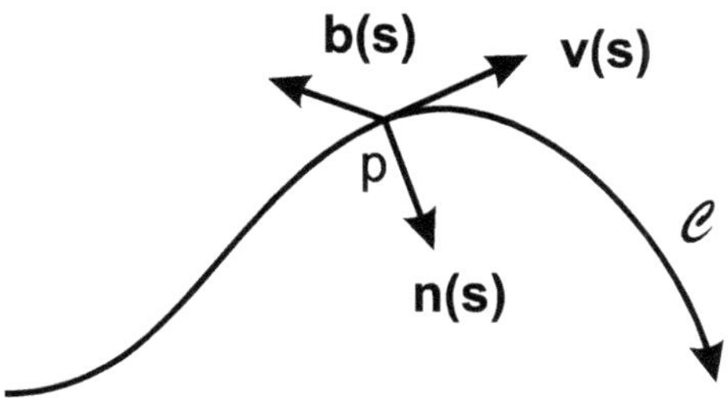

Figure 4.4. The tangent, normal, and binormal vectors $\boldsymbol{v}(s)$, $\boldsymbol{n}(s)$, and $\boldsymbol{b}(s)$ form an orthonormal triad at each point p along the curve $\mathcal{C}$. Here, the curve is drawn in the plane of the page, so that $\boldsymbol{v}$ and $\boldsymbol{n}$ are in the plane of the page and the binormal vector is perpendicular to the page, pointing away from the reader.

The derivative of $\boldsymbol{n}$ cannot have a component along the direction of $\boldsymbol{n}$ (otherwise the length would be changing and it would not remain a unit vector), so it can only have $\boldsymbol{b}$ and $\boldsymbol{v}$ components. Similarly, $\boldsymbol{b}'$ can't have a $\boldsymbol{b}$ component, and in fact can only have an $\boldsymbol{n}$ component. So the derivatives of the three unit vectors are given by the **Frenet formulas**:

$$\boldsymbol{v}' = k\boldsymbol{n} \tag{4.28}$$

$$\boldsymbol{n}' = -k\boldsymbol{v} - \tau\boldsymbol{b} \tag{4.29}$$

$$\boldsymbol{b}' = \tau\boldsymbol{n}, \tag{4.30}$$

where τ is called the **torsion**. The curvature k measures how fast the tangent vector bends in the $\boldsymbol{v}$–$\boldsymbol{n}$ plane (the **osculating plane**), while the torsion τ measures how fast the osculating plane itself is rotating as you move along the curve. For completeness, the $\boldsymbol{v}$–$\boldsymbol{b}$ plane is called the **rectifying plane** and the $\boldsymbol{n}$-$\boldsymbol{b}$ plane is called the **normal plane**.

4.3.2 Two-dimensions and beyond

A **homeomorphism** is a map that is one-to-one, onto (surjective), and continuous, with a continuous inverse. Homeomorphisms represent continuous deformations of manifolds, such as bending, stretching, and twisting; but ripping new holes or filling in old holes are not allowed. A **regular surface** is a two-dimensional manifold S such that there is a mapping $\boldsymbol{f}$ from an open neighborhood V of each point to an open neighborhood U in $\mathbb{R}^2$, such that

1. $\boldsymbol{f}$ is differentiable,
2. $\boldsymbol{f}$ is a homeomorphism, and
3. $\boldsymbol{f}_*$ is one-to-one.

f_* is called the **differential** or **pushforward** of $\boldsymbol{f}$, and is sometimes denoted $d\boldsymbol{f}$. It maps vectors tangent to $\mathcal{S}$ to vectors in $\mathbb{R}^n$. The differential map serves as a higher dimensional generalization of the parameter derivative $\boldsymbol{f}'$ in the one-dimensional case.

The set of all vectors tangent to two-dimensional surface $\mathcal{S}$ at a point $p \in \mathcal{S}$ is the tangent space TV_p of V at p. Similarly, there is a tangent space at the image of p: $TU_{f(p)}$. Let $\{x^1, x^2\}$ be coordinates on $U \subset \mathbb{R}^2$ and $\{y^1, y^2\}$ be coordinates on $V \subset \mathcal{S}$. Since the differential f_* takes vectors (which can be thought of as column matrices) to vectors, f_* can be thought of as a matrix. We can define coordinate basis vector fields $\partial/\partial y^\alpha$ and $\partial/\partial x^\mu$ on V and U, respectively, where $\alpha, \mu \in \{1, 2\}$. Vectors tangent to $\mathcal{S}$ and $\mathbb{R}^2$ can then be written in component form as $\boldsymbol{w} = w^\alpha(\partial/\partial y^\alpha)$ and $\boldsymbol{r} = r^\mu(\partial/\partial x^\mu)$. Then the differential f_* takes $\boldsymbol{w}$ to the new vector $\boldsymbol{r}$:

$$\boldsymbol{r} = f_*(\boldsymbol{w}), \tag{4.31}$$

$$\begin{pmatrix} r^1 \\ r^2 \end{pmatrix} = \begin{pmatrix} \frac{\partial x^1}{\partial y^1} & \frac{\partial x^1}{\partial y^2} \\ \frac{\partial x^2}{\partial y^1} & \frac{\partial x^2}{\partial y^2} \end{pmatrix} \begin{pmatrix} w^1 \\ w^2 \end{pmatrix}. \tag{4.32}$$

More compactly,

$$r^\mu = w^\alpha \frac{\partial x^\mu}{\partial y^\alpha}. \tag{4.33}$$

In other words, this is just the formula for a change of variables in a vector, between the coordinates of $U \subset \mathbb{R}^2$ and the coordinates of $V \subset \mathcal{S}$.

If the differential map is surjective (onto) at a point, the image of the initial surface's tangent space fills all the dimensions of the image tangent space, rather than collapsing to a lower dimensional subspace, so the point is a **regular point**. Any point where this is not true is called a **critical point**. At critical points, $\mathrm{Det}(df) = 0$, providing a higher dimensional generalization of the condition $df/dx = 0$ for maxima and minima of one-dimensional functions.

The tangent spaces of $\mathcal{S}$ are two-dimensional like the original surface. The basic idea can be generalized to differentials of arbitrary differentiable maps between any two manifolds: the dimensions of the matrix just change to match those of the spaces. For example, the differential of a map from an m-dimensional manifold to an n-dimensional manifold will be a matrix of m columns and n rows. Also notice that when the spaces involved are Euclidean spaces, $\mathbb{R}^n$, there are simplifications since $\mathbb{R}^n$ is isomorphic to its own tangent space; there is no need to distinguish between $\mathbb{R}^n$ and $T\mathbb{R}^n$.

As was true for curves, surfaces also have an important vector associated to each point which is *not* in the tangent space. This is the **normal vector**, $\boldsymbol{N}$, which is perpendicular to the surface and to its tangent space at each point. If $\boldsymbol{e}_1$ and $\boldsymbol{e}_2$ are a pair of basis vectors tangent to the surface, then the unit normal vector can be defined by the product:

$$\boldsymbol{N}(p) = \frac{\boldsymbol{e}_1 \times \boldsymbol{e}_2}{|\boldsymbol{e}_1 \times \boldsymbol{e}_2|}. \tag{4.34}$$

Notice that there are two possible orders in which you could multiply the vectors on top, and interchanging them introduces a minus sign. These two orderings are referred to as two **orientations** for the surface. If you can choose a consistent orientation at all points (like on a sphere or a torus), the surface is called **orientable**; if not (like on a Möbius strip, figure 4.5), it is **non-orientable**.

The normal vector field defines a mapping, $\boldsymbol{N}: U \to T\mathbb{R}^n \simeq \mathbb{R}^n$, associating a normal vector to each point of U. If the surface is orientable, then this can be extended from U to the full surface $\mathcal{S}$. Furthermore, since $\boldsymbol{N}$ is a unit vector, its tip always lies on a sphere of unit radius around the surface point. Denoting the two-dimensional sphere by S^2, we can therefore be more specific, and say that on an orientable surface the normal vectors define a map $\boldsymbol{N}: \mathcal{S} \to S^2$. This is the **Gauss map**.

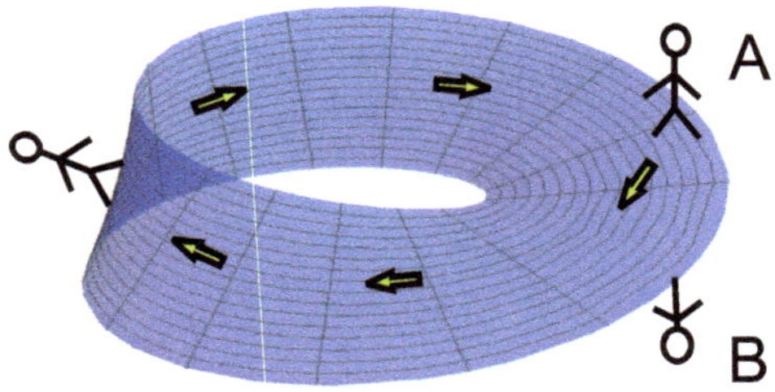

Figure 4.5. Möbius strip. A walker starting at point A and completing one circuit of the strip will have opposite orientation when he returns to the same point (B). Any attempt at covering the strip with a continuous field of normal vectors will end up giving a field that is double-valued. The strip is therefore non-orientable. Similarly, a minimum of two coordinate patches are required to consistently cover the strip without becoming double-valued.

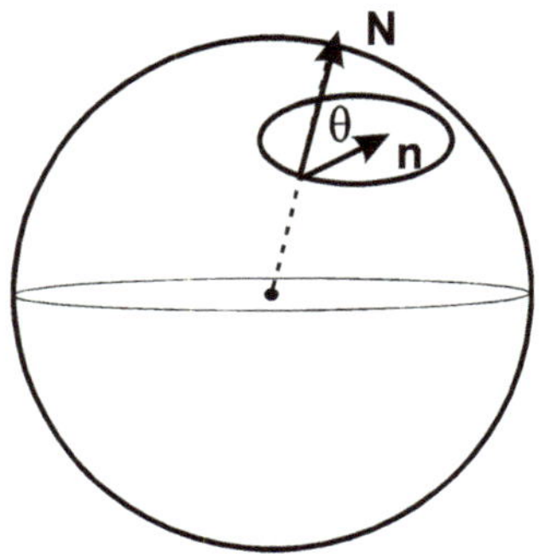

Figure 4.6. Any point on a sphere has a normal vector $\boldsymbol{N}$ pointing either toward or away from the center of the sphere, depending on the orientation of the surface. The normal $\boldsymbol{n}$ to a curve drawn on the surface, however, can have components both tangent and normal to the sphere. In the case of the circle drawn here, $\boldsymbol{n}$ points toward the center of the circle.

Suppose a curve $\alpha(s)$ is drawn on the surface, k being its curvature. There are now two unit normals: $\boldsymbol{n}$ and $\boldsymbol{N}$ are, respectively, the normal vectors to the curve and to the surface. On a flat surface, these are always perpendicular to each other, since $\boldsymbol{n}$ will be tangent to the surface. On curved surfaces this is not true: for example, think of a circle drawn on a sphere, where an outward-pointing normal $\boldsymbol{N}$ will be directed away from the center of the *sphere*, but $\boldsymbol{n}$ will point toward the center of the *circle* and so may have a component tangential the sphere, as in figure 4.6.

Let θ be the angle between $\boldsymbol{n}$ and $\boldsymbol{N}$, so that $\cos\theta = \langle \boldsymbol{n}, \boldsymbol{N} \rangle$. Here, the bracket represents the inner product defined by the metric. Then the **normal curvature** k_n of the surface is the projection of $\boldsymbol{v}' = k\boldsymbol{n}$ onto the direction normal to the surface; in other words,

$$k_n = \langle \boldsymbol{v}', \boldsymbol{N} \rangle = k \cos\theta. \tag{4.35}$$

k_n is unchanged if you travel backward along the curve, but flips direction if you reverse the orientation of the surface.

The **principal curvatures**, k_1 and k_2, at a point are the minimum and maximum values of the normal curvature, extremized over all curves in $\mathcal{S}$ passing through the point. These are eigenvalues of the differential map $d\boldsymbol{N}$. The corresponding unit eigenvectors $\hat{e}_1$ and $\hat{e}_2$ give the **principal directions**. On a cylinder the principle directions are parallel to the axis (minimum curvature) and perpendicular to the axis (maximum curvature). On a sphere, the normal curvature is constant so that all directions at any point are principle directions.

Given a unit vector $\boldsymbol{v}$ tangent to the surface, it can always be decomposed into components along the two principle directions,

$$v = \hat{e}_1 \cos\theta + \hat{e}_2 \sin\theta. \tag{4.36}$$

Then the **normal curvature** along any curve that has $\boldsymbol{v}$ as a tangent vector at that point is given by

$$k_n = k_1 \cos^2\theta + k_2 \sin^2\theta. \tag{4.37}$$

The **Gaussian curvature** K and the **mean curvature** H are, respectively, the determinant and half the trace of $-d\boldsymbol{N}$, where $d\boldsymbol{N}$ is the differential of the Gauss map:

$$K = \mathrm{Det}(-d\boldsymbol{N}) = k_1 k_2 \qquad \text{(Gaussian curvature)} \tag{4.38}$$

$$H = \frac{1}{2}\mathrm{Tr}(-d\boldsymbol{N}) = -\frac{1}{2}(k_1 + k_2) \qquad \text{(Mean curvature)} \tag{4.39}$$

The principal curvatures are then

$$k_{1,2} = H \pm \sqrt{H^2 - K}. \tag{4.40}$$

Positive Gaussian curvature means that two locally perpendicular curves on the surface bend in the same direction (for example, on a sphere), while negative Gaussian curvature means they bend in opposite directions (as at a saddle point). Vanishing Gaussian curvature means that at least one direction is flat (for example, at any point on a cylinder or plane).

Going from two dimensions to higher dimensions, the mean, Gaussian, and normal curvatures (which are just numbers) will be generalized to tensor quantities like the Riemann and Ricci curvature tensors, which are most conveniently defined directly from the metric tensor.

An equivalent picture of curvature is as a lack of commutativity of small displacements; on a curved surface, carrying out two consecutive displacements in different directions will give different results if implemented in opposite order. The reader can easily verify this: on a ball or a spherical balloon, start at the equator and draw a line segment going due east, then a segment of the same length going north. Going back to the starting point, repeat the same steps in opposite order. You will find that you end up at a different final point each time. The difference will be proportional to the curvature.

One further formula will be mentioned without proof for curvature on a surface. Let $\hat{e}_{1,2}$ be a pair of basis vector fields on a surface with metric g and let ∂_1 and ∂_2 be derivatives along the directions of the basis vectors. On a flat space, these derivatives will always commute: $[\partial_1, \partial_2]f(\boldsymbol{r}) \equiv (\partial_1\partial_2 - \partial_2\partial_1)f(\boldsymbol{r}) = 0$, at every point and for every function f on the surface. On a curved space, this will no longer necessarily be true, so the lack of commutativity of derivatives can be taken as a measure of curvature. In fact, denoting the inner product by brackets, the Gaussian curvature can be written as

$$K = \frac{1}{\det(g)}\langle[\nabla_1, \nabla_2]\hat{e}_1, \hat{e}_2\rangle, \tag{4.41}$$

where ∇represents the covariant derivative, which is defined in the next section. This view of curvature as lack of commutativity of derivatives or small displacements is the basis for forming more general ideas of curvature in higher dimensions.

4.4 Connections and covariant derivatives

Suppose that you are driving on a long, straight stretch of highway in your expensive new sports car. Intellectually, you know that the highway is not really straight, but that it in fact follows the curving of the Earth. However, your senses tell you the path is straight. Not only does it look straight, but you also feel no feeling of following a curved path: there is no sensation of centripetal acceleration perpendicular to the ground. This is partly because the radius of the Earth is so large, but it is also in part due to the fact that we tend to take the Earth's surface as the reference by which we measure motion. Only when the curve of the road bends horizontally, tangent to the surface on which you are traveling, does the curving become completely apparent. Essentially, being confined to a two-dimensional surface, motions within the surface are clearly visible, but motions of the confining surface itself perpendicular to its tangent plane are largely invisible from our viewpoint. The curved motion along the 'straight' road is, however, clearly visible to an alien observer preparing his plan for world domination while orbiting on a satellite. Such an external observer, floating well outside the confines of the planet's surface, clearly sees the road bending to follow the Earth's surface and finds it obvious that our motion is fully three-dimensional.

Ordinary space derivatives like d/dx or ∇will describe how this field appears to the outside observer on the satellite; this three-dimensional view of the system is the **extrinsic** view, which can see how the 2D earth and the 1D road curve within a larger three-dimensional space. But we would like to be able to also give an **intrinsic** description of curvature and motion, relying only on quantities that can be measured by an observer within the lower-dimensional space, to whom the extra dimensions outside the curve or surface are invisible. This leads us the ideas of connection and covariant derivative.

Consider a curve $\gamma(s)$ on a surface $\mathcal{S}$, and some vector field $v(s)$ defined along the curve and tangent to $\mathcal{S}$; our car's velocity, for example. Although $v(s)$ is tangent to the surface, its derivative $v' = dv/ds = (\partial v_j/\partial x_i)(dx_i/ds)\hat{e}_j$ may not be, since the curving of the surface itself may introduce a component of v' perpendicular to $\mathcal{S}$. The perceived change in v, as viewed by the observer moving along the curve, will then be the projection of this non-tangential vector v' back into the tangent plane. This is the **covariant derivative along the curve**:

$$\nabla_\gamma v(s) = v' - (\text{normal component of } v') \tag{4.42}$$

$$= v' - \langle v', N\rangle N, \tag{4.43}$$

where N is the unit normal vector of the surface. Thinking of a vector $w = w^i\partial_i$ tangent to the curve as representing displacements by amounts w^i along the curves represented by the coordinate lines, we use the linearity of derivatives to define the **covariant derivative in the direction of** w,

$$\nabla_w v = \nabla_{w^i\partial_i} v = w^i \nabla_i v, \tag{4.44}$$

where $\nabla_i \equiv \nabla_{\partial_i}$ is the covariant derivative along the relevant coordinate line. The derivative $\nabla_i v$ is the *perceived* rate of change of v in the ith coordinate direction relative to an observer confined to S.

Of course, the normal vector is not an intrinsic quantity to the surface, so we want to write this in a more useful form. To that end, look at the action of the covariant derivative on one of the basis vectors. Again, this action must be linear, so the result must itself be a linear combination of the various basis vectors. Therefore, we can define a set of quantities Γ^k_{ij} as the coefficients in this linear combination,

$$\nabla_j \hat{e}_i = \Gamma^k_{ij} \hat{e}_k. \tag{4.45}$$

In the context of studying tangent vector fields on manifolds, the Γ^k_{ij} are called **Christoffel symbols**. They are special cases of **connection coefficients**, used to describe tangent vectors on fiber bundles (section 4.5).

The ∇_i are normally thought of as components of a covariant derivative operator, $\nabla = \nabla_i dx^i$. This is a map from rank (k,l) to rank $(k, l+1)$ tensors. (The components of a (k,l) tensor have k upper indices and l lower indices.) Given any two tensors T and S of any rank and any complex numbers a and b, the covariant derivative is required to satisfy the following conditions:

(1) **Linearity:** $\nabla_i(aT + bS) = a\nabla_i T + b\nabla_i S$
(2) **Leibnitz rule:** $\nabla_i(T \otimes S) = (\nabla_i T) \otimes S + T \otimes (\nabla_i S)$ (This is a straightforward generalization of the product rule of differential calculus.)
(3) The act of applying the covariant derivative should commute with the act of contracting a pair of indices: $\nabla_i(T^j_{jk}) = (\nabla_i T)^j_{jk}$. (This is equivalent to saying that the Kronecker delta used to contract the indices is constant: $\nabla_i \delta_{jk} = 0$.)
(4) The covariant derivative reduces to the ordinary partial derivative when acting on functions (0-forms): $\nabla_i f(x) = \partial_i f(x)$.

The action of the covariant derivative on vectors and one-forms is given by

$$\nabla_i V^j = \partial_i V^j + \Gamma^j_{ik} V^k \tag{4.46}$$

$$\nabla_i \omega_j = \partial_i \omega_j - \Gamma^k_{ij} \omega_k, \tag{4.47}$$

with the metric on the space used to raise and lower derivatives. The coefficients Γ^k_{ij} are initially thought of as the ^k_j components of a matrix Γ_i, with one such matrix for each value of i, but the distinction between the three indices starts becoming blurred as they get raised, lowered, and permuted in various formulas. Since an arbitrary tensor can be built out of vectors and one-forms, equations (4.46) and (4.47) determine the action of the covariant derivative on all tensors of any rank. (In passing, it should be mentioned that the connection matrix Γ_i is not a true tensor, since it does not transform correctly under coordinate changes.)

When we discuss electromagnetic or gauge connections at the end of this chapter, the corresponding connections analogous to Γ^k_{ij} will be given by the gauge field components, A_i. There are two fewer indices because the missing indices only have a single value each (they exist on a one-dimensional space), and are therefore unnecessary. There is only a single value because in the gauge field case, these indices do not represent directions in physical space but directions in the internal group space. This is a one-dimensional group, called $U(1)$, whose elements form a circle; see section 4.6.

The connection coefficients that can be defined on a given manifold are not unique, and manifolds with different connections will have different properties with respect to vector fields and inner products. For example, in general relativity two additional requirements are imposed (metric compatibility and vanishing torsion) in order to uniquely pin down the connection coefficients.

Given two points p and q in flat Euclidean space, there is no problem in comparing vectors at those points, because all points in the space are identical, as are their tangent spaces. But on a curved space, the comparison of vectors at different points is more ambiguous, since there is no unique relationship between the two tangent spaces $T(p)$ and $T(q)$. The additional structure provided by connections and covariant derivatives is used to remove this ambiguity: a vector at p can be parallel transported along a curve to q, providing a vector in $T(q)$ that is the 'most parallel' vector possible to the original one. This new vector may then be unambiguously compared to other vectors in the same tangent space by means of the inner product at q.

The 'straightest possible curve' γ on a surface is that for which the vector tangent to the curve is covariantly constant, i.e. the covariant derivative of the tangent vector with respect to itself vanishes:

$$\nabla_v v = 0, \tag{4.48}$$

where $v(s) = d\gamma(s)/ds$. Such a curve is called a **geodesic**. Similarly, a vector field $\boldsymbol{w}(s)$ is said to be **parallel transported** along a curve with tangent vector field $v(s)$ if

$$\nabla_v \boldsymbol{w} = 0. \tag{4.49}$$

Under parallel transport, the component of $\boldsymbol{w}$ normal to the surface may vary along the curve, but the component of $\boldsymbol{w}(s)$ tangent to the surface remains at a fixed angle with the vector $v(s)$ as $\boldsymbol{w}$ is dragged along the curve.

This now provides a mean of comparing vectors at different points on the manifold: to compare vector $\boldsymbol{w}$ at point p to vector $\boldsymbol{u}$ at point q, simply parallel transport $\boldsymbol{w}$ along some curve to point p, and then measure its angle from $\boldsymbol{u}$. Note that this procedure depends on the existence of a connection (or covariant derivative). It also depends on the curve used: parallel transport of the same vector along different curves can lead to different results. This is provides another view of curvature: the difference between the parallel transported vectors along different paths will be proportional to the curvature.

Not all curves are geodesics. To find a geodesic, one must solve the **geodesic equation** $\nabla_v v = 0$; a vector field $v(t)$ that solves this equation is then tangent to a geodesic and integrating it, starting from initial vector $v(0)$, gives the geodesic curve itself. In terms of components, the geodesic equation becomes

$$\frac{d}{dt}v^i + \Gamma^i_{\sigma\rho}\frac{dx^\sigma}{dt}v^\rho = 0, \tag{4.50}$$

where t is the parameter along the curve.

Geodesics have a number of special properties, which we won't prove here. These include:

(1) The path of minimal length between two points is always a geodesic.
(2) Given any point p on a manifold and any tangent vector v at that point, a geodesic can always be found that passes through p and whose tangent at p is v.
(3) If an inner product is defined, then at any point of a d-dimensional manifold a set of d geodesics can always be chosen whose tangent vectors are orthogonal to each other, and in a sufficiently small neighborhood of that point the geodesics will define axes for an orthonormal coordinate system.

On simple surfaces, the geodesic curves are often obvious (figure 4.7). On a flat plane, geodesics are simply straight lines. On a sphere, they are the great circles, i.e. circles which pass through any pair of antipodal points. On a cylinder, lines parallel to the axis and circles and helices that wrap around the cylinder are examples of geodesics, while on a torus the simplest geodesics include two groups of circles: those that circle the hollow part of the torus and those that circle the 'donut hole'.

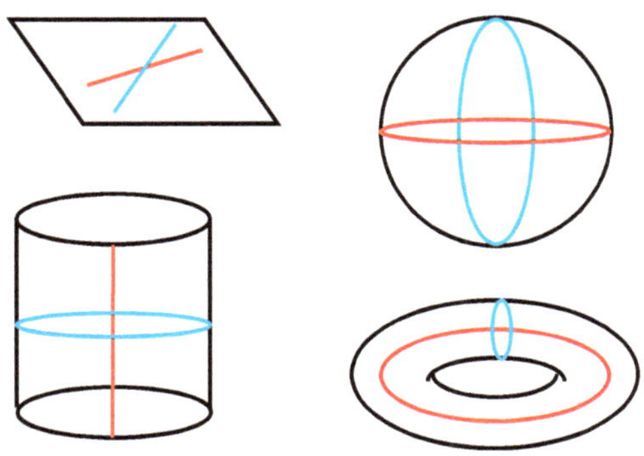

Figure 4.7. Examples of geodesics in two dimensions. On a plane, any straight line is a geodesic, while any great circle is a geodesic for a sphere. The geodesics of a cylinder include straight lines parallel to the axis and circles running azimuthally around the cylinder's hole. For tori, the geodesics include sets of circles enclosing, respectively, the large and small holes of the torus. On both the torus and the cylinder, there also exist geodesics that spiral around the surface.

Geodesics often have direct physical significance. For example in general relativity, the path that a free particle follows will be a geodesic of the curved spacetime. In optics, if a material has a spatially varying index of refraction, then the gradient of the index can be thought of as inducing an effective curvature and a connection; the light rays then follow paths that are geodesic with respect to this connection.

In this section, the curvature and connection of a *manifold M* have been discussed. In the next section, these ideas will be generalized to the more abstract setting of *fiber bundles*. These bundles are the natural mathematical setting for describing physical fields, and it will be seen that the electromagnetic field (and other gauge fields) serve as connections for particular types of bundles.

4.5 Fiber bundles

This section gives a brief introduction to the idea of fiber bundles, which have come to play a large role in our description of the fundamental process of nature, particularly in gauge theories that describe fundamental forces such as electromagnetism, gravity, and nuclear forces. The fiber bundle picture will give a natural interpretation to the idea of geometric phases in chapter 9.

This section gives a very general picture of fiber bundle structures (figure 4.8). Much of what is given here is more general than needed for the rest of the book. Some readers may prefer to quickly read just the first few paragraphs to get the basic definition of a fiber bundle, then move on the next section where we specialize from the general case to the simpler case relevant to electromagnetism and optics.

A **fiber bundle** (E, π, M, F, G) consists of the following parts: E is a topological space called the **total space**; F is a topological space known as the **fiber**; the manifold M is called the **base manifold**; π is a projection map $\pi: E \to M$; and finally, G is the group of homeomorphisms of F to itself and is called the **structure group** of the bundle. The transformations in the structure group move points around within the fiber, while leaving the base point fixed. The projection 'collapses' the copy of F located at point x (denoted F_x), down onto the base point x, $\pi: F_x \to x$. Conversely, the fiber is the inverse image of the base point: $F_x = \pi^{-1}(x)$. Fiber bundles formalize and generalize the notion of 'gluing' a copy of space F,

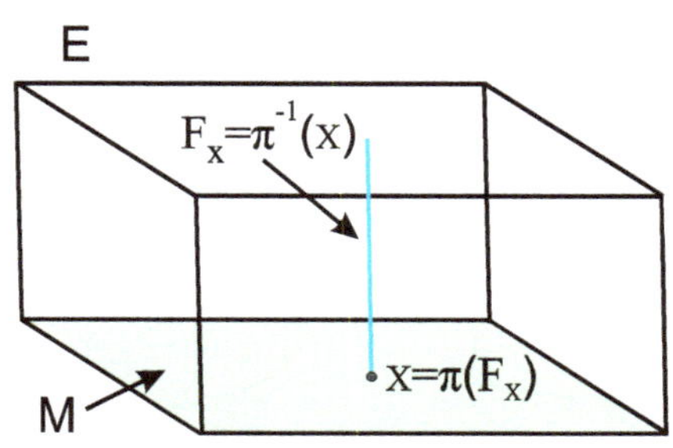

Figure 4.8. The fiber bundle E consists of a base manifold M, with another space called the fiber attached to each point of M. In this image, the two-dimensional base has a one-dimensional fiber F at each point, with the horizontal directions representing the base and the vertical direction representing the fiber. One fiber, F_x, is drawn explicitly, given by the inverse image of the projection map π at point $x \in M$.

representing internal variables of a particle or field, onto each point x of a spacetime M. Bundles are often referred to by the name of their total space, E, for short. More explicitly, a bundle is often denoted by the sequence

$$F \to E \to M. \tag{4.51}$$

Let M be covered by a collection of open sets $\{U_\alpha\}$, with an associated set of differentiable mappings ϕ_α: $U_\alpha \times F \to \pi^{-1}(U_\alpha)$, such that $\pi \circ \phi_\alpha(x, f) = x$, for all $(x, f) \in U_\alpha \times F$, where $\circ$denotes composition of mappings or functions. These maps are called **local trivializations**; if it is possible for M to be covered by a single set and map with these properties, then the bundle is said to be **trivial** and the bundle is then just the direct product of M and F. On the overlaps of the $\{U_\alpha\}$ we define a set of **transition functions**

$$g_{\alpha\beta} = \phi_\alpha \circ \phi_\beta^{-1}: (U_\alpha \cap U_\beta) \times F \to (U_\alpha \cap U_\beta) \times F, \tag{4.52}$$

satisfying the consistency conditions

$$g_{\alpha\alpha} = 1 \qquad g_{\alpha\beta} = g_{\beta\alpha}^{-1} \qquad g_{\alpha\beta} g_{\beta\gamma} = g_{\alpha\gamma}. \tag{4.53}$$

A **local section** is a continuous map s_α: $U_\alpha \to E$ such that $\pi \circ s(x) = x$, for all $x \in U_\alpha$. Similarly a *global* section, or simply a **section** is a similar map defined on all of M. The space of all sections of a bundle E over M is denoted $\Gamma(M, E)$. Sections are essentially slices or cross-sections of the fiber bundle. A global section can only be defined if the bundle is trivial (figure 4.9).

Some important special cases of fiber bundles include the following:

- A **principle bundle** is formed by taking F to be given by the structure group G itself. G then acts on the fiber by right multiplication: $g \in G$: $h \to h \cdot g$,

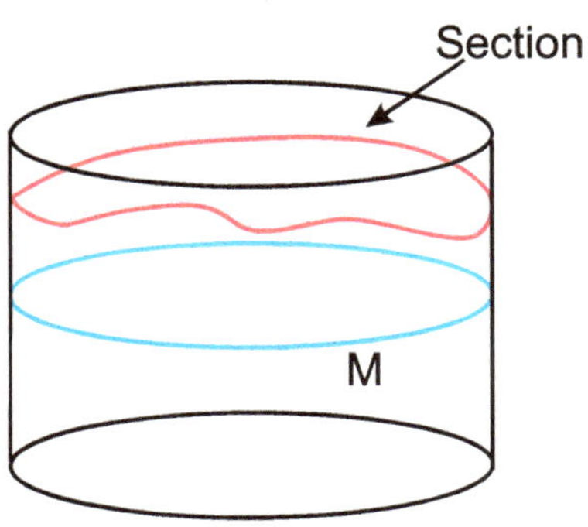

Figure 4.9. The red curve represents a section in the bundle. Here the base space is the blue circle, and the fibers are vertical intervals I perpendicular to the base at each point, making the bundle topologically equivalent to the surface of a cylinder. The image of the projection operator π applied to the section covers the entire base. Since there is a single section for the entire bundle, the bundle is trivial and can be globally represented as a product of the circle and the interval, $S^1 \times I$. More generally, it may be impossible to construct a global section. For example, if the fibers going around the circle were given a twist before reconnecting to the starting point, the result would be nontrivial bundle with the same base and fiber, but whose total space would be a Möbius strip.

where the dot represents the group multiplication. A generic fiber bundle (E, π, M, F, G) can be shown to be trivial if the related principle bundle $P = (E, G)$ has a global section.

- In a **vector bundle**, the fiber is given by a vector space, on which G acts via some linear matrix representation. The fiber dimension of a vector bundle is referred to as the bundle's **rank**.
- A **line bundle** is a vector bundle with a fiber of a single (real or complex) dimension.
- The **tangent bundle** TM and **cotangent bundle** T^*M of manifold M. The fibers are the sets of vectors or covectors at a given point, and the structure group is $GL(m, \mathbb{R})$, the group of $m \times m$ real, invertible matrices, where m is the dimension of M. Similarly, the spaces of tensors of other types can be viewed as the direct product of powers of the tangent and cotangent bundles.

Consider a principle bundle P with base M, and let TP be the tangent bundle over P. Given the fiber F_x of P over point $x \in M$, the tangent bundle TF of the fiber can be used to define a vertical direction for TP: anything tangent to F is declared to be vertical. At any point in P, the vertical tangent space can then be identified with the Lie algebra $\mathcal{G}$ of the structure group G: for every $\xi \in \mathcal{G}$, we can uniquely identify a vertical vector $v_\xi \in T_pP$. But since there is no intrinsic notion of orthogonality on the bundle, it is necessary to add some extra structure in order to specify the horizontal direction. This structure is usually given in the form of a **connection one-form**, ω. When a *vertical* vector v_ξ is inserted into this structure, the result is the corresponding Lie algebra element: $\omega(v_\xi) = \xi$. Then a vector v is defined to be *horizontal* if

$$\omega(v) = 0. \tag{4.54}$$

In other words, the connection form projects vectors onto their vertical component and annihilates the horizontal part. This is a generalization of the connection or covariant on a manifold in section 4.4, which projected onto the part of a vector that was in the direction of the normal vector $\boldsymbol{N}$. Notice that when we insert a vector into ω we annihilate the vector (as is done by a one-form) and replace it with a Lie algebra element; so ω is a Lie-algebra-valued one-form.

Given a curve $c(t)$ in M, we can now define the **horizontal lift** of this curve to be the curve $\tilde{c}(t)$ in P such that for every t:

(i) $\pi\tilde{c}(t) = c(t)$, and
(ii) the tangent vector $(d/dt)\tilde{c}(t)$ is horizontal.

In fact, given $c(t)$ in M with $c(0) = x$ and a point p contained in the fiber $\pi^{-1}(x) = F_x$, there exists a unique curve $\tilde{c}(t)$ through P which is both a horizontal lift of $c(t)$ and which satisfies $\tilde{c}(0) = p$. This result provides us with a notion of **parallel transport** in the bundle: a vector at point p of the fiber is parallel transported along the curve $c(t)$ in M by pushing it along the horizontal curve $\tilde{c}(t)$ starting at p using the connection, just as on a manifold. Section of vector bundles give states of quantum particles, or in other words the section represents a wavefunction.

Since points in the fiber may represent vectors (describing internal variables such as spin or angular momentum), the covariant derivative and the gauge potential will act on sections to transform them into new sections, with the gauge field acting as a matrix (an **endomorphism**) on the vector-valued fiber. Given a section s of a principle G-bundle P, a physical gauge field $A(x)$ with gauge group G lifts via the inverse of the bundle projection to define a connection one-form ω along section s:

$$\omega(p) \rightarrow A(x) = A_i(x)dx^i, \qquad p \in \pi^{-1}(x) \tag{4.55}$$

Here, A is a one-form on the base manifold, defined locally within a given open set, and A_i is in general matrix-valued for non-Abelian gauge fields. The choice of section corresponds to a choice of gauge: if instead of $s(x)$ we had chosen a different section (a different gauge),

$$s'(x) = s(x)g(x), \tag{4.56}$$

where at each point $g(x)$ is a Lie group element, then the field transforms as

$$A' = g^{-1}Ag + g^{-1}dg = A_i' \, dx^i. \tag{4.57}$$

In components, this becomes the transformation law for gauge fields:

$$A_i' = g^{-1}A_i g + g^{-1}\partial_i g. \tag{4.58}$$

If the bundle is nontrivial, travelling from one open set to another requires switching to a different section, i.e. doing a gauge transformation. So nontrivial bundles have can have no single global choice of gauge. The group element g serves to move the section up and down the fibers.

The **covariant exterior derivative** corresponding to the connection is defined as follows. If σ is a q-form and $X_1, X_2, \cdots, X_{q+1}$ are vectors, then

$$D\sigma(X_1, \ldots, X_{q+1}) = d\sigma(X_1^H, \ldots, X_{q+1}^H), \tag{4.59}$$

where X_i^H is the horizontal part of X_i and d is the exterior derivative; i.e D gives the horizontal part of the derivative.

The **curvature one-form** of a principle bundle is given by the covariant derivative of the connection one-form:

$$\Omega = D\omega. \tag{4.60}$$

This is a Lie algebra-valued two-form. Unlike the connection, it transforms homogeneously under gauge transformations: $g\colon \Omega \rightarrow g^{-1}\Omega g$.

As with the connection form, we can project the curvature form onto the base to get a Lie algebra-valued two-form F on M. The projection of equation (4.60) becomes

$$F(x) = dA + A \wedge A = \frac{1}{2}F_{ij}dx^i \wedge dx^j, \tag{4.61}$$

with components,

$$F_{ij} = \partial_i A_j - \partial_j A_i + [A_i, A_j]. \tag{4.62}$$

It can be easily checked that the curvature simply measures the commutativity of covariant displacements in different directions,

$$F_{ij} = -i\left[D_i, D_j\right], \tag{4.63}$$

where $D_i v^a = \partial_i v^a - iA^a_{ib} v^b$.

Starting with a projection π: $P \to M$, we can define the fiber $\pi^{-1}(x)$ over each point $x \in M$. Given a closed curve $C(t)$ in the base manifold M such that $C(0) = C(2\pi) = x_0$, we can lift the curve using π^{-1} to construct a lifted curve $\tilde{C}(t)$ in the bundle. To make this curve unique, require that it be horizontal; i.e., the tangent vectors v^μ to $\tilde{C}(t)$ are required to be parallel transported along $\tilde{C}(t)$:

$$D_\mu v^a = 0. \tag{4.64}$$

However, the lifted curve may not be closed: $\tilde{C}(0)$ and $\tilde{C}(1)$ may be different elements of the same fiber $\pi^{-1}(x_0)$ (figure 4.10). Let $z(0)$ and $z(1)$ be the coordinates in the fiber $\pi^{-1}(x_0)$ of the initial and final points. Then they will differ by the action of some group element g acting on the right,

$$z(1) = z(0)g. \tag{4.65}$$

g is called the **holonomy** of the curve C. The curve thus defines a mapping of the fiber to itself. The holonomies of all the curves passing through the base point x_0 form a subgroup of the structure group G, called the **holonomy group** of the connection at the given point, $H(A, x_0)$. For a connected manifold, the holonomy groups at different base points are isomorphic, so the dependence on x_0 may be dropped; in this case, the holonomy group of the connection can be denoted $H(A)$, with elements denoted $\Phi(C, A)$. Since the fibers over each point of M are all isomorphic, note that holonomy can also be defined along curves that are not closed.

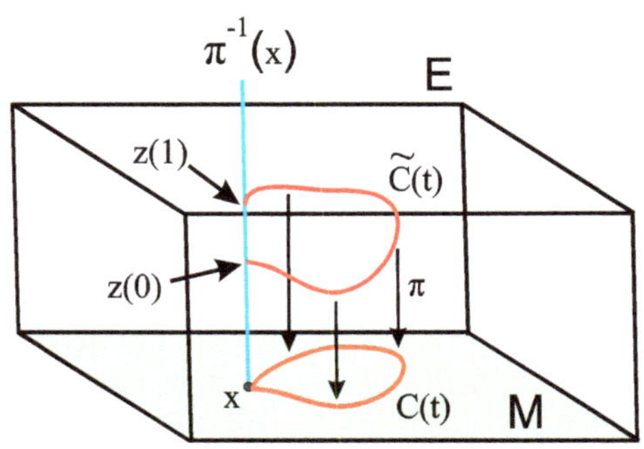

Figure 4.10. The closed loop $C(t)$ in the base is horizontally lifted to a curve $\tilde{C}(t)$ in the bundle. When C returns to its starting point, $\tilde{C}$ must return to the same fiber, but not necessarily the same point in the fiber. The group element g that translates the initial point in the fiber $z(0)$ to the final point $z(1)$ defines the holonomy element associated with the closed curve C.

The subgroup $H^0(A)$ restricted to contractible loops is called the **restricted holonomy group** or **local holonomy group**. If M is simply connected, then $H(A) = H^0(A)$. In any event, there is a homomorphism of the fundamental group M, $\pi_1(M) \to H(A)/H^0(A)$, in which a homotopy class $[C]$ is mapped to the coset $\Phi(C, A)H^0(A)$. (Recall that the coset space means that all elements of $H(A)$ that are connected by elements of $H^0(A)$ are identified with each other.) A connection is flat if and only if $H^0(A)$ is trivial.

Given a gauge theory with connection A, the **Dirac phase factor** acquired around a closed curve C is

$$\Phi(C, A) = e^{iq\int_{x0}^{x1} A_\mu(x)dx^\mu} = e^{iq\int_{x0}^{x1} F_{\mu\nu}dx^\mu \wedge dx^\nu}, \tag{4.66}$$

where q is a coupling constant. (In the non-Abelian case, path ordering of the integral is necessary but we won't worry about that here.) Under a gauge transformation $A \to A'$, the phase factor around a closed loop is gauge invariant, $\Phi(A) = \Phi(A')$, and $\Phi(A)$ is an element of the holonomy group. For non-Abelian groups, the **Wilson loop** of closed curve C is defined to be

$$W(C) = \langle Tr\Phi(C, A)\rangle, \tag{4.67}$$

where the trace is over the group indices. Unlike the gauge field itself, the Wilson loops are gauge-invariant and physically observable. In nuclear physics, the Wilson loop serves as an order parameter: its value distinguishes between confined and unconfined phases of quark-gluon plasmas. In the Abelian case of electromagnetism, the holonomy around loops is the geometric phase factor discussed in chapter 9.

4.6 Connection and curvature in electromagnetism and optics

The last section gave a general and abstract description of fiber bundles, but here we wish to extract the portions that are most important to the case of interest for this book: the $U(1)$ principle bundle that describes electromagnetic fields. This will be much simpler than the general case for two reasons: (i) the fiber is one-dimensional (it is essentially a circle), and (ii) electric fields are Abelian or commutative: $\left[A_\mu, A_\nu\right] = 0$. Physically, the Abelian nature of the electromagnetic field means that photons do not interact directly with each other. (It should be kept in mind that they can still interact indirectly. For example, in non-linear optics multiple photons can interact with the crystal lattice of a solid; the lattice then mediates an effective interaction between the photons.)

The group $U(1)$ is the group of one-dimensional unitary transformations; it acts by multiplication by complex numbers of unit modulus. In other words, the elements of the group are phase factors of the form $e^{i\theta}$ with real θ. These factors lie on the unit circle in the complex plane. The fields therefore live on a $U(1)$ principle bundle whose base is spacetime M and whose fiber is a circle representing the $U(1)$ factors.

Consider a particle of charge q moving through an electromagnetic field. Local $U(1)$ invariance means that at each point $\boldsymbol{x}$ physically measurable quantities should be invariant under the transformation

$$\psi(\boldsymbol{x}) \rightarrow \psi'(\boldsymbol{x}) = e^{iq\theta(x)}\psi(\boldsymbol{x}), \tag{4.68}$$

where θ is an arbitrary function. In other words, we can freely choose any value for θ, and the value we choose at one point can be completely independent of the choice at other points. If the particle is moved from $\boldsymbol{x}$ to $\boldsymbol{x} + d\boldsymbol{x}$, the change in wavefunction, including the phase factor, is

$$\psi'(\boldsymbol{x} + d\boldsymbol{x}) + \psi'(\boldsymbol{x}) = \left[\partial_\mu\psi'(\boldsymbol{x}) - iq\big(\partial_\mu\theta(\boldsymbol{x})\big)\psi'(\boldsymbol{x})\right]dx^\mu \tag{4.69}$$

$$\equiv D_\mu\psi'(\boldsymbol{x})dx^\mu, \tag{4.70}$$

where the covariant derivative is

$$D_\mu = \partial_\mu + iqA_\mu, \tag{4.71}$$

and the connection $A_\mu(\boldsymbol{x}) \equiv \partial_\mu\theta(\boldsymbol{x})$ is the gauge field. θ represents the location on the $U(1)$ fiber, and A_μ measures the rate at which our choice of θ changes from point to point. We are free to make a local gauge transformation, i.e. a position-dependent rotation of the phase angle,

$$\theta(\boldsymbol{x}) \rightarrow \theta(\boldsymbol{x}) + \lambda(\boldsymbol{x}), \tag{4.72}$$

or equivalently,

$$\psi(\boldsymbol{x}) \rightarrow e^{iq\lambda(x)}\psi(\boldsymbol{x}) \equiv g(\boldsymbol{x})\psi(\boldsymbol{x}), \tag{4.73}$$

where $g(\boldsymbol{x}) = e^{iq\lambda(x)}$ is an element of the $U(1)$ group. This introduces λ-dependent terms into the Lagrangian. In order to make the theory gauge-invariant, these terms must be canceled. It turns out that this requires the field A_μ to also be transformed according to

$$A_\mu(\boldsymbol{x}) \rightarrow A_\mu(\boldsymbol{x}) - \partial_\mu\lambda(x), \tag{4.74}$$

which, for the form of g above, is the Abelian version of equation (4.58). The particle wavefunction and its derivative then both transform the same way under the group,

$$\psi \rightarrow g\psi \tag{4.75}$$

$$D_\mu\psi \rightarrow g\big(D_\mu\psi\big), \tag{4.76}$$

which ensures the invariance of the Lagrangian. Since the fields commute, equations (4.62) and (4.63) for the curvature reduce to the familiar form,

$$F_{\mu\nu} = -\frac{i}{q}\left[D_\mu, D_\nu\right] = \partial_\mu A_\nu - \partial_\nu A_\mu. \tag{4.77}$$

Gauge transformations of ψ are compensated by gauge transformations of the field A_μ, which correspond to taking different sections of the $U(1)$ fiber bundle. If units are taken so that $\hbar = 1$, then the structure here is identical to that described for the electromagnetic field in chapter 2. Geometrically, the electromagnetic field represents a connection on a $U(1)$ fiber bundle, and its field tensor $F_{\mu\nu}$ is the curvature of the bundle. The nontrivial phase factors gained in moving a particle around a closed loop represent the holonomy, or nonclosure of the path when lifted from the base to the fiber. Examples of holonomy in optics will be seen in chapter 9.

4.7 The Hopf fibration and polarization

One additional important example of a fiber bundle that comes up often in physics is the **Hopf bundle**. Recall that $SU(2)$ is the group of 2×2 special unitary matrices, the unitary matrices with determinant equal to $+1$. An element s of $SU(2)$ can be viewed as a two-by-two complex matrix of the form:

$$s = \begin{pmatrix} \alpha & -\beta^* \\ \beta & \alpha^* \end{pmatrix}, \tag{4.78}$$

where $|\alpha|^2 + |\beta|^2 = 1$. The reader can readily verify that

$$s^\dagger s = 1 \quad \text{and} \quad \mathrm{Det}(s) = 1. \tag{4.79}$$

Notice that the pair of complex numbers α and β define a unit vector in $\mathbb{R}^4$; in other words, a point on the surface of the three-dimensional unit sphere S^3. We therefore have an isomorphism between the unitary group and the three-sphere, $SU(2) \sim S^3$.

Starting from $s \in SU(2)$, we now define a three-dimensional vector $\boldsymbol{x} \in \mathbb{R}^3$ whose components are given by

$$x^i = \frac{1}{2}\mathrm{Tr}(\sigma_i s \sigma_3 s^{-1}), \tag{4.80}$$

where σ_i are the three Pauli matrices. This vector is a unit vector, $|\boldsymbol{x}|^2 = 1$, which means that the mapping π: $s \to \boldsymbol{x}$ is a map between spheres,

$$\pi: S^3 \to S^2. \tag{4.81}$$

This is a two-to-one mapping, since s and $-s$ both project onto the same $\boldsymbol{x} \in S^2$.

Explicitly, the mapping between $\boldsymbol{x}$ and s is given by:

$$\alpha = \cos|\boldsymbol{x}| + ix^3 \sin|\boldsymbol{x}| \tag{4.82}$$

$$\beta = (ix^1 - x^2)\sin|x|, \tag{4.83}$$

or equivalently,

$$x^1 = \mathrm{Re}(2\alpha^*\beta) \tag{4.84}$$

$$x^2 = \mathrm{Im}(2\alpha^*\beta) \tag{4.85}$$

$$x^3 = |\alpha|^2 - |\beta|^2. \tag{4.86}$$

Notice further that the points of S^2 are in a one-to-one correspondence with the elements of the group of rotations in three dimensions, the special orthogonal group $SO(3)$. This can be seen by choosing an arbitrary unit vector $\hat{n}$, which singles out a reference point on S^2, or equivalently, a reference direction in $\mathbb{R}^3$. Then any other point on S^2 (any other unit vector) specifies a different direction in $\mathbb{R}^3$, which can be obtained by a rotation (an element of $SO(3)$) of $\hat{n}$. Thus we have the isomorphism $S^2 \sim SO(3)$, and the projection of equation (4.81) can equally well be written as a mapping between groups

$$\pi\colon SU(2) \to SO(3). \tag{4.87}$$

The mapping π is in fact a fiber bundle projection, projecting the total space $SU(2) \sim S^3$ onto the base $SO(3) \sim S^2$. Two elements s and s' of the fiber are related by a structure group element $g = s^{-1}s'$, which can be easily shown to take the form

$$g = s^{-1}s' = e^{i\theta\sigma_3}, \tag{4.88}$$

for some real number θ. We see therefore that the fiber F and the structure group G form a circle in the complex plane, which in turn is isomorphic to the group U(1): $F \sim S^1 \sim U(1)$. We therefore have a principle fiber bundle given by the **Hopf fibration**

$$S^1 \to S^3 \to S^2, \tag{4.89}$$

or

$$U(1) \to SU(2) \to SO(3). \tag{4.90}$$

For an example of the Hopf fibration in optics, consider polarization. We can define a unit vector $\boldsymbol{x}$ on the Poincaré sphere S^2 with coordinates $x^i = S_i/|\boldsymbol{S}|$, for $i = 1, 2, 3$, where S_i are the Stokes parameters of section 2.4. Assuming light is propagating along the z-axis, we can parameterize the nonzero components of the electric field by

$$E_i = A_i \cos\left(\omega t - kz + \phi_i\right), \qquad i = 1, 2. \tag{4.91}$$

For convenience, we can normalize the amplitude, $A_1^2 + A_2^2 = 1$, which allows the A_i to be parameterized by an angular variable,

$$A_1 = \cos\frac{\theta}{2}, \qquad A_2 = \sin\frac{\theta}{2}. \tag{4.92}$$

Defining the relative phase $\delta\phi = \phi_2 - \phi_1$, the components of $\boldsymbol{x}$ can then be written as

$$x_1 = 2A_1A_2 \cos\delta\phi = \sin\theta\cos\delta\phi \tag{4.93}$$

$$x_2 = 2A_1A_2 \sin\delta\phi = \sin\theta\sin\delta\phi \tag{4.94}$$

$$x_3 = A_1^2 - A_2^2 = \cos\theta. \tag{4.95}$$

The point $\boldsymbol{x}$ corresponds to spinor $s = \begin{pmatrix} z_1 \\ z_2 \end{pmatrix} \in SU(2) \sim S^3$, which in the same notation as above can be written as

$$z_1 = \cos\frac{\theta}{2}e^{i\phi_1} \tag{4.96}$$

$$z_2 = \sin\frac{\theta}{2}e^{i\phi_2}. \tag{4.97}$$

The complexified electric field components can then be written in terms of the spinor components as $E_i = z_i e^{i(\omega t - kz)}$, and the map π: $s \to \boldsymbol{x}$ is precisely the projection of Hopf fibration.

References

[1] do Carmo M P 1982 *Riemannian Geometry* (Boston, MA: Birkhäuser)
[2] Pressley A N 2010 *Elementary Differential Geometry* II edn (London: Springer)
[3] Lovett S T 2010 *Differential Geometry of Manifolds* (Natick, MA: A. K. Peters/CRC Press)
[4] Guillemin V and Pollack A 1974 *Differential Topology* (Englewood Cliffs, NJ: Prentice-Hall)
[5] Hirsch M W 1976 *Differential Topology* (New York: Springer)
[6] Steenrod N 1951 *The Topology of Fibre Bundles* (Princeton NJ: Princeton University Press)
[7] Nakahara M 2003 *Geometry, Topology and Physics II edn* (Boca Raton, FL: CRC Press)
[8] Nash C 1992 *Differential Topology and Quantum Field Theory* (London: Academic)
[9] Nash C and Sen S 2013 *Topology and Geometry for Physicists* (Mineola, NY: Dover)
[10] Schutz B F 1980 *Geometrical Methods of Mathematical Physics* (Cambridge : Cambridge University Press)
[11] Moriyasu K 1983 *An Elementary Primer For Gauge Theory* (Singapore: World Scientific)
[12] Chan H M and Tsun T S 1993 *Some Elementary Gauge Theory Concepts* (Singapore: World Scientific)

IOP Publishing

Topology in Optics (Second Edition)
Tying light in knots
David S Simon

Chapter 5

Topological invariants

Topological invariants are quantities which are unchanged by homeomorphisms and which capture some information about the topological properties of a space. They usually take discrete values, which can be normalized to take on integer values. Topological invariants encode information about the global structure of a space, but can often be defined in terms of local quantities such as curvature. Recall from chapter 3 that many homotopy groups are isomorphic to the set of integers, or to some subset of the integers: each homotopy class can then be labeled by some appropriately chosen integer-valued topological invariant.

Integer-valued topological invariants have become common in quantum field theory, solid state physics, optics, and other areas of physics. In this context they are often referred to as **topological quantum numbers** or **topological charges**, since they are conserved quantities whose values are quantized to integers (in appropriately chosen units) similar to standard quantum numbers like the principle quantum number of the hydrogen atom.

In this chapter, integer-valued topological invariants will be discussed as mathematical objects. Physical realizations of these quantities will appear repeatedly in the coming chapters.

5.1 Euler characteristic

For a manifold M, the **Euler–Poincaré characteristic** (often simply called the Euler characteristic) can be defined via triangulations. Every sufficiently well-behaved space can be given a triangulation; we describe the idea for two dimensions, but a similar process can be carried out in higher dimensions.

Cover a two-dimensional surface S by a network of line segments or curved arcs called **edges**, which join up at a set of points, called **vertices**. Together, the lines and edges form a graph on the surface (figure 5.1). Arrange the vertices and edges so that the entire surface is covered by a set of triangles enclosed by the edges, with the vertices lying at the corners of the triangles. These triangles are called faces. Let E, F, and

doi:10.1088/978-0-7503-3471-6ch5

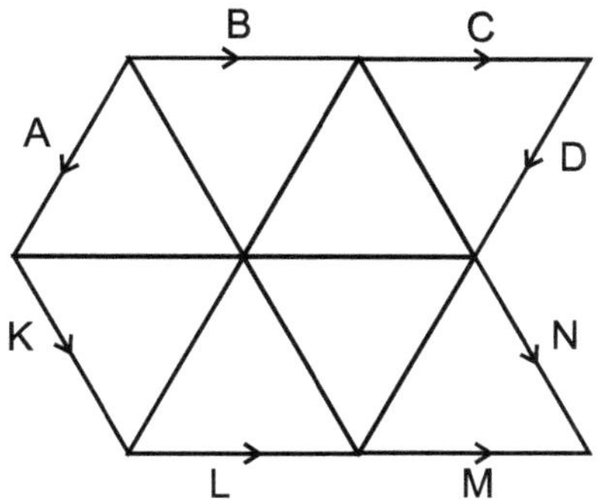

Figure 5.1. Triangulation of a two-dimensional surface. A closed surface can be formed by identifying edges with each other, so that the arrows on the identified edges are pointing in the same direction. If we identify pairs of edges with each other according to $B = L$, $C = M$, $A = D$, and $K = N$, with arrows as shown above, then the resulting surface is a torus. However, if we add a twist by reversing the arrows on edges D and N (and identifying $A = N$, $K = D$), the result is a non-orientable surface called a Klein bottle. Either way, we have $E = 12$, $V = 4$, and $F = 8$, so that $\chi = 4 - 12 + 8 = 0$.

V represent the number of edges, faces, and vertices in this triangulation. Then the Euler characteristic for the surface is defined by the formula

$$\chi(\mathcal{S}) = F - E + V. \tag{5.1}$$

Notice that the sign alternates with dimension: the even dimensional vertices and faces come with plus signs, the odd dimensional edges come with a minus sign. For a more general manifold M, the pattern will continue with higher dimensional objects added into the alternating sum: $\chi(M) = \sum_n (-1)^n b_n$, where b_n is the nth **Betti number** (the rank of the nth homology group).

The Euler characteristic is in fact independent of the triangulation: any triangulation that can be drawn on the surface will give the same result. This is easy enough to see through examples. Starting from a surface with some triangulation already given on it, and consider what happens when the triangulation is changed by adding an additional vertex and corresponding edges, as in figure 5.2. When we add the new vertex, this requires adding three new edges, which in turn replaces the original face inside which the vertex was added by three smaller faces. So

$$V \rightarrow V + 1, \qquad E \rightarrow E + 3, \qquad F \rightarrow F + 2, \tag{5.2}$$

which results in

$$\chi = F - E + V \rightarrow (F + 2) - (E + 3) + (V + 1) = F - E + V = \chi. \tag{5.3}$$

In other words, the Euler characteristic is unchanged. Any other changes that one tries to make in the triangulation will similarly leave χ invariant. So χ is a property solely of the underlying space, not of the triangulation used to describe it.

The Euler characteristic can be described in another, equivalent way. From a topological point of view, any two-dimensional orientable surface can be obtained from either a sphere or a plane by adding a set of handles to it. (We assume here that the manifolds have no boundary.) For example, a torus is obtained by adding a

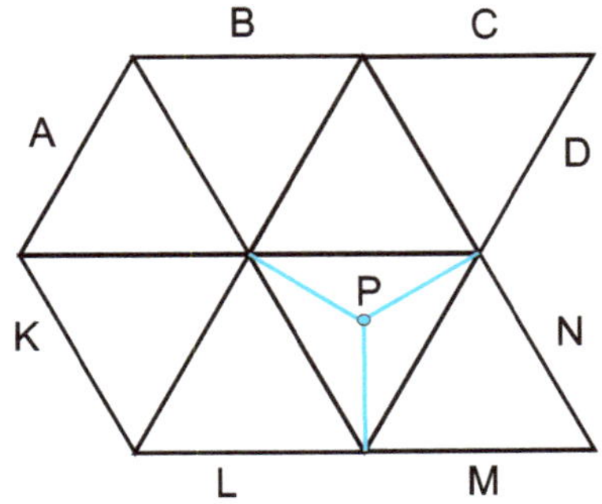

Figure 5.2. A new triangulation obtained from the one of the previous figure by adding an additional vertex at P. This forces the addition of three new edges (in blue), which in turn subdivides the original face into three smaller faces. These alterations leave χ unchanged.

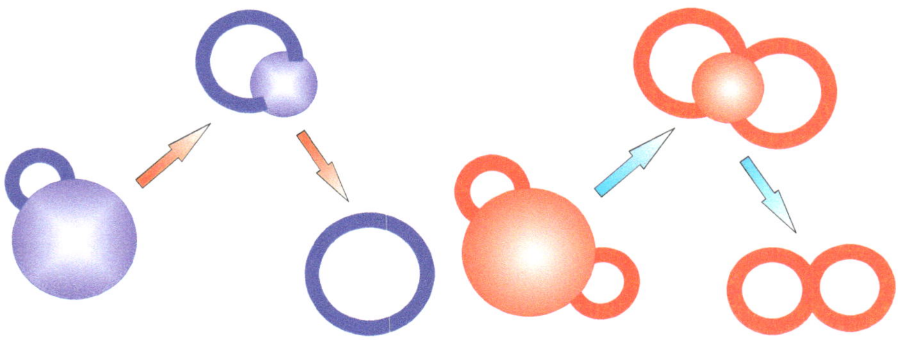

Figure 5.3. Single-holed (a) and double-holed (b) tori can be continuously deformed into two-dimensional spheres with handles (tubes) attached. Each handle is attached by removing two disks from the sphere and sewing the ends of handle onto the boundary circles formed by removing the disks. By using a triangulation of the surface, it should be easy to see that each attached handle changes the Euler characteristic by −2.

handle to a sphere, as in figure 5.3. The number of such handles is called the **genus** g of the surface. Surfaces formed by adding handles to a sphere are compact, those formed from planes are non-compact. The Euler characteristic of a manifold M can then be defined by means of the genus:

$$\chi(M) = 2 - 2g. \tag{5.4}$$

In two dimensions, any compact boundaryless surface of genus $g = 0$ can be continuously deformed into a sphere, while any such surface of genus 1 can be deformed into a torus (or equivalently, into a sphere with one handle attached). Figures 5.3(a) and 5.3(b) illustrate the formation of single- and double-holed tori by adding handles to spheres. If the surface has a boundary, the definition of the Euler characteristic is generalized to

$$\chi(M) = 2 - 2g - b, \tag{5.5}$$

where b is the number of connected components that make up the boundary.

We have assumed above that the surface is orientable. If it is *non-orientable*, the handles are given a twist (reversing the orientation of one end, as in the triangulation of the Klein bottle) before attachment.

The generalized **Poincaré conjecture** asserts that the situation in two dimensions generalizes to arbitrary dimension: that any closed, compact manifold of any dimension is homeomorphic to an n-sphere S^n with handles (twisted if unorientable or untwisted otherwise) attached. First conjectured by Poincaré in 1904, it was not until 1961 that American mathematician Stephen Smale proved that the conjecture is true for all $n > 4$. This was followed by a proof for $n = 4$ in 1982 by Michael Freedman, and finally for $n = 3$ in a series of papers in 2002 and 2003 by Russian mathematician Grigori Perelman.

The Euler characteristic is closely related to the average curvature of a manifold. In two dimensions, the connection is given by the famous **Gauss–Bonnet theorem**:

$$\int_M K\, dA = 2\pi\chi(M), \tag{5.6}$$

where K is the Gaussian curvature (chapter 4). This theorem relates a geometric object (curvature) to a topological object (Euler characteristic or genus). The Gauss–Bonnet theorem has a number of generalizations, including the **Chern–Gauss–Bonnet theorem** (on even-dimensional, compact, boundaryless manifolds), the **Riemann–Roch theorem** (on Riemann surfaces), and the **Atiyah–Singer index theorem** (for differential operators on compact manifolds).

If the manifold M has a boundary ∂M, then a term must be added to take into account the curvature of the boundary:

$$\int_M K\, dA + \int_{\partial M} k_g\, ds = 2\pi\chi(M), \tag{5.7}$$

where s is arc-length.

The Gaussian curvature and the Euler characteristic can be positive, negative, or zero. Representative examples of each case include:

(1) **Sphere** ($g = 0$): $\chi = 2$ and $K > 0$ everywhere.

(2) **Torus** ($g = 1$): Now $\chi = 0$. The single-hole, two-dimensional torus has regions of both positive and negative Gaussian curvature, with an average curvature of zero.

(3) **Two-handled torus** ($g = 2$): Now $\chi = -2$ and the average curvature is negative, $K < 0$.

In regions of $K > 0$, initially parallel geodesics converge toward each other; think of positive curvature as being similar to the action of a converging lens, bending light rays toward each other. Meanwhile on regions of $K = 0$ and $K < 0$ parallel geodesics remain parallel or diverge, respectively. Free particle motion on surfaces of negative curvature exhibit *sensitive dependence on initial conditions*: no matter how close two particles start to each other, they always move apart. This is a hallmark of chaotic behavior. A surface with constant negative Gaussian curvature is called a **pseudosphere**.

On a surface, it is possible to draw a geodesic triangle, a triangle whose sides are segments of three intersecting geodesics. Let ϕ_i be the interior angles at the three corners of the triangle. Then, from the Gauss–Bonnet theorem one can show that

$$\sum_{i=1}^{3} \phi_i = \int K\, dA + \pi. \tag{5.8}$$

Therefore, it follows that

$$K > 0 \leftrightarrow \sum_{i=1}^{3} \phi_i > \pi \qquad \text{example: sphere} \tag{5.9}$$

$$K = 0 \leftrightarrow \sum_{i=1}^{3} \phi_i = \pi \qquad \text{example: cylinder or plane} \tag{5.10}$$

$$K < 0 \leftrightarrow \sum_{i=1}^{3} \phi_i < \pi \qquad \text{example: pseudosphere} \tag{5.11}$$

5.2 Winding number

Consider a two-dimensional surface $\mathcal{S}$ punctured at point P (figure 5.4). In other words, $\mathcal{S}$ has a hole due to the removal of point P from the space. The remaining space is denoted $\mathcal{S}' = \mathcal{S} - \{P\}$. Take some starting point $x_0 \in \mathcal{S}'$, and consider some closed path $\gamma: I \to \mathcal{S}'$, where $I = [0, 1]$ is the unit interval, with $\gamma(1) = \gamma(0)$. Drawing a line L from P to x_0, we may define the angle of any point $\gamma(s)$ on the curve from L. θ may be thought of as a coordinate on the circle, S^1, and as one progresses along $\gamma(s)$, this angle may not be single-valued, since θ and $\theta + 2\pi$ represent the same point. We therefore unwrap the circle to form a line $\mathbb{R}$, as in figure 5.5, allowing θ to have any angle from 0 to ∞. (In the terminology of fiber bundles (chapter 4), the new $\mathbb{R}$-valued function $\tilde{\theta}(s)$ is a *lift* of the multivalued function $\theta(s)$. θ and $\tilde{\theta}$ are strictly speaking different functions, but henceforth we will simply denote both functions as θ.)

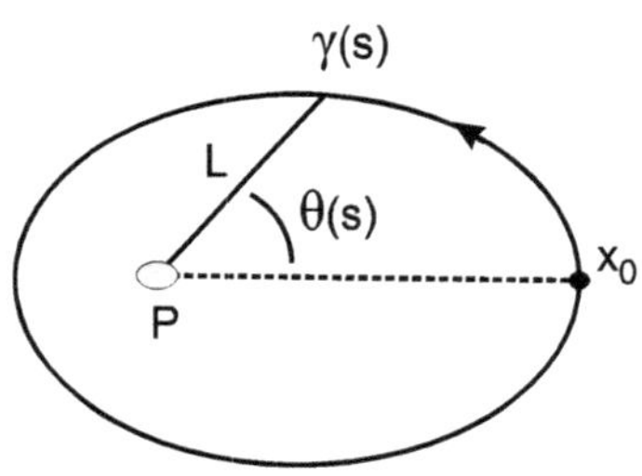

Figure 5.4. A surface with a puncture at P. A line can be drawn to the basepoint, as a reference axis from with angle θ can be defined. As a closed curve $\gamma(s)$ is traced out, the angle $\theta(s)$ equals a multiple of 2π upon traversing a complete circuit. This integer winding number labels the homotopy class of the closed curve.

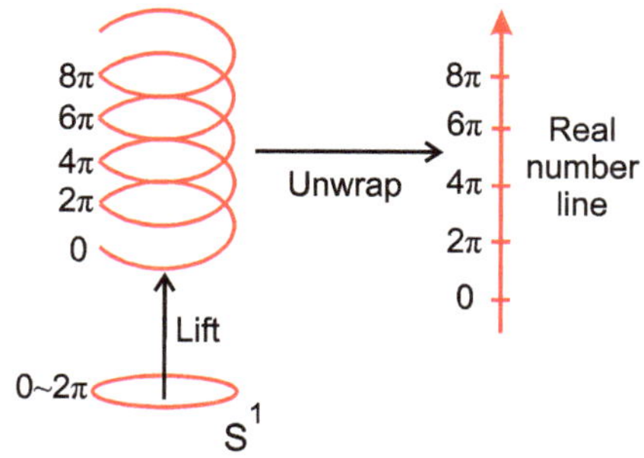

Figure 5.5. The lift of the circle S^1 is the set of real numbers $\mathbb{R}$. The circle is unwound to remove the periodicity.

As closed loop $\gamma(s)$ completes its circuit from $s = 0$ to $s = 1$, the angle $\theta(s)$ evolves from an initial value $\theta(0)$ to final value $\theta(1)$. The winding number of path γ about point P is then defined to be

$$n(\gamma, P) = \frac{1}{2\pi}[\theta(1) - \theta(0)] = \frac{1}{2\pi}\int_0^1 \frac{d\theta}{ds}ds. \tag{5.12}$$

The winding number is clearly an integer, since when the curve returns to its starting point it must end up at an angle that differs from its initial value by an integer multiple of 2π. Equally clear is the intuitive meaning of n: it counts the number of times the curve encloses P before returning to its starting point, with $n > 0$ for counterclockwise windings and $n < 0$ for clockwise.

An important theorem provides the connection between winding number and homotopy: Two loops γ_1 and γ_2 in $\mathcal{S}'$ are homotopic to each other if and only if they have the same winding number about P. In other words, the winding number can be used to label the first homotopy class.

This theorem can be generalized to more complicated situations. For example, $\mathcal{S}$ may be punctured at multiple points, P_j, for $j = 1,\ldots,n$. Every curve will then have multiple winding numbers: there will be a winding number $n_j(\gamma) = n(\gamma, P_j)$ about each puncture. Then two loops in the punctured space will be homotopic if and only if their complete set of winding numbers $\{n_1, \ldots, n_n\}$ are the same.

5.3 Index of zero points of vector fields

Let U be an open set of a topological space and let V be a vector field on U. A **zero** of V is a point $P \in U$ where all the components of V vanish, $V = 0$. A zero at P is **isolated** if there is a neighborhood of P that contains no other zeros. Let $\mathcal{Z}$ denote the set of zeros and $U' = U - \mathcal{Z}$ be the complementary set of points in U where the field is nonvanishing.

Consider a closed loop γ that encloses isolated zero P a single time in the counterclockwise direction. We can define the vector field along the curve, $V_\gamma(s) \equiv V(\gamma(s))$. The **index**, $\mathcal{I}_P(V) = \mathrm{Index}_P(V)$, of the vector field V about zero

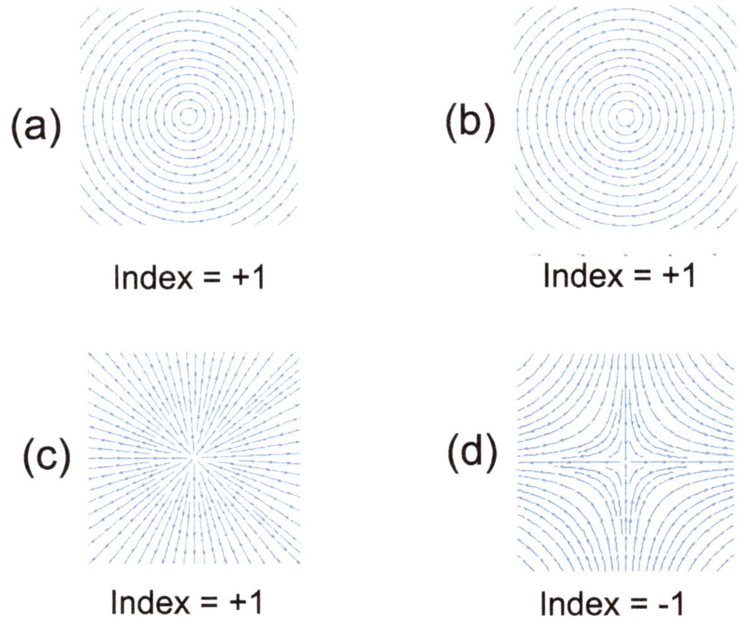

Figure 5.6. The index of a zero point of a vector field is the number of times a vector rotates as it is taken counterclockwise around a closed path encircling the zero. For fields circulating in either direction around the zero in (a) or (b), the index is one. For a radial field like a source (c) or a sink (not shown), the index is also one. But for a saddle point (d), the index is -1.

P is given by the number of rotations of the vector field $V_\gamma(s)$ about P as one circulates counterclockwise along $\gamma(s)$. Examples are shown in figure 5.6. If the rotation of the vector is counterclockwise, the index is positive; for clockwise rotations it is negative. If P is the only singular point, then $V_\gamma(s)$ is independent of the chosen path (as long as the path completes one circuit of P), so the dependence on γ is usually dropped. If there are multiple singular points, then the index of the field is the sum of the indices at all of the singular points, $\mathcal{I} = \sum_P \mathcal{I}_P$.

Vector fields can always be related to differential operators (see chapter 4), so that index theorems, such as the Atiyah–Singer theorem or Riemann–Roch theorem which involve various types of generalized indices, provide linkages between (i) differential equations on a space, (ii) the possible vector fields on the space, (iii) and the topology of the space (see section 8). In chapter 6 several topological numbers relevant to optics will be defined which can be viewed as indices of the type defined above.

Unlike the winding number, the index is a topological invariant attached to a vector field on the manifold, not to the manifold itself. However, the topology of the manifold restricts the possible values of index the field may have on that space. For example, in two dimensions, we have the **Poincaré–Hopf theorem**: Given a vector field $V(r)$ on a surface $\mathcal{S}$ with only isolated zeros (given by the set $\mathcal{Z}$), then the sum of the indices about the zeros will equal the Euler characteristic of the surface:

$$\sum_{z\in\mathcal{Z}} \mathcal{I}_P(V) = \chi(\mathcal{S}). \tag{5.13}$$

In chapter 6 we discuss vortices in optical fields; these vortices will contain zeros of vector fields at their centers. The Poincaré-Hopf theorem guarantees that these vortices must be created and annihilated in pairs in order to conserve the Euler characteristic of the space in which the vortices exist.

5.4 Chern numbers

Consider a compact two-dimensional manifold M and a map to the two-sphere, $\boldsymbol{\phi}: M \to S^2$. In other words, for $\boldsymbol{r} \in M$, $\boldsymbol{\phi}(\boldsymbol{r}) = (\phi_x(\boldsymbol{r}), \phi_y(\boldsymbol{r}))$, with $\phi_x^2 + \phi_y^2 = 1$, where $\boldsymbol{r} = (x, y)$ in some coordinate system on M. Something that looks suspiciously like a curvature tensor can be defined, $F_{ij} = \partial_i\phi_j - \partial_j\phi_i$, and from this curvature we may construct a topological invariant:

$$c_1 = \frac{1}{2\pi}\int_M dx\, dy\, F_{xy}. \tag{5.14}$$

This is the **first Chern number**, and it takes values in the integers. Notice that equation (5.14) looks very reminiscent of the expression for the Euler characteristic as an integral over Gaussian curvature (the Gauss–Bonnet theorem); the Chern number is in fact one of many generalizations of the Euler characteristic.

Even for non-compact spaces, the integral defining the Chern number may converge if the curvature vanishes at large distances. It is often convenient to replace non-compact two-dimensional surfaces by equivalent compact surfaces. For example, the plane $M = \mathbb{R}^2$ can be **compactified** into a two-sphere by collapsing the circle at infinity to a point, effectively sewing together the 'edges' at infinity. Topologically, this means identifying M with a two-sphere, $M \sim S^2$, so that the mapping ϕ takes spheres to spheres, ϕ: $S^2 \to S^2$. In the notation of chapter 3, the compactification can be written $S^2 \sim M/\partial M$. c_1 essentially counts the number of times the first sphere wraps around the second, and its integer values will characterize the second homotopy class, $\pi_2(S^2)$.

In physical applications, $\boldsymbol{\phi}(\boldsymbol{r})$ will often represent a point on the Bloch or Poincaré sphere, x, y will be momentum components, and the two dimensional manifold M will be a torus T^2 representing a Brillouin zone. In this context, c_1 is often referred to as the **TKNN invariant** (after Thouless, Kohomoto, Nightingale, and den Nijs [1]); it turns out to be important in the quantum Hall effect, and it will make an appearance in chapter 10.

c_1 can also be thought of as the integral $c_1 = \int_M \omega$ over M of a two-form $\omega = 1/4\pi F_{\mu\nu}dx^\mu \wedge dx^\nu$, called the **first Chern class**. Although c_1 is the only Chern number we will be concerned with, it is the first of an infinite sequence of Chern numbers defined in even numbers of dimensions. The nth Chern number is given in $2n$ dimensions by the integral over the base of a $2n$-form on a bundle with a complex fiber:

$$c_n = \frac{1}{n!}\left(\frac{1}{4\pi}\right)^n \int \epsilon^{\mu_1\nu_1\mu_2\nu_2\ldots\mu_n\nu_n} F_{\mu_1\nu_1} F_{\mu_2\nu_2} \cdots F_{\mu_n\nu_n} dx^1 \wedge dx^2 \wedge \cdots \wedge dx^{2n}$$
$$= \frac{1}{n!}\int_M Tr(\omega^n), \tag{5.15}$$

where the trace is over the degrees of freedom described by the fiber. $\epsilon^{\mu_1\nu_1\mu_2\nu_2\cdots\mu_n\nu_n}$ is the completely antisymmetric tensor that vanishes if any of the indices are equal, equals +1 if the indices form an even permutation of $\{1, 2, \ldots, n\}$, and equals −1 for odd permutations. The $2n$-form being integrated over is called the nth Chern class. These higher Chern classes and Chern numbers won't be needed here but do come up in other areas of physics, such as quantum field theory. For more details, see [2, 3].

5.5 Pontrjagin index

The second and third homotopy classes of the two-spheres are both labeled by integers:

$$\pi_2(S^2) = \pi_3(S^2) = \mathbb{Z}. \tag{5.16}$$

These labels are known, respectively, as the Pontrjagin and Hopf indices, and are discussed, respectively, in this section and the next.

Consider a map $\boldsymbol{\phi}: I \times I \rightarrow S^2$. This could, for example, represent a scalar field with internal degrees of freedom forming a three-dimensional vector lying on the unit sphere in the internal space. Such internal variables could be the three 'color' degrees of freedom in the theory of strong nuclear interactions or the components of a polarization vector in optics. The condition that the field be a unit vector in the internal space is written as $\boldsymbol{\phi} \cdot \boldsymbol{\phi} = 1$ or $\phi_a\phi^a = 1$, where the dot product and the latin indices are in the internal target space of the mapping. Components in the original square $I \times I$ will be given Greek indices. Let x^μ with $\mu = 1, 2$ be coordinates on the unit square $I \times I$. We can compactify the square by collapsing all the points on the boundary to a single point, thereby identifying the square with the two-sphere. Therefore, $\boldsymbol{\phi}$ can be viewed as a map between spheres: $\boldsymbol{\phi}: S^2 \rightarrow S^2$, thereby defining a homotopy class in $\pi_2(S^2)$.

Define a quantity

$$Q = \frac{1}{8\pi}\int_{I\times I} \epsilon^{\mu\nu}\boldsymbol{\phi} \cdot (\partial_\mu\boldsymbol{\phi} \times \partial_\nu\boldsymbol{\phi})\, dx^1\, dx^2 = \frac{1}{4\pi}\int_{I\times I} A, \tag{5.17}$$

where

$$A = \boldsymbol{\phi} \cdot \left(\frac{\partial\boldsymbol{\phi}}{\partial x^1} \times \frac{\partial\boldsymbol{\phi}}{\partial x^2}\right) dx^1 \wedge dx^2 = \frac{1}{2}\epsilon_{abc}\phi^a\partial_\mu\phi^b\partial_\nu\phi^c\, d^\mu \wedge dx^\nu, \tag{5.18}$$

where again the cross product is acting on the internal indices. Q is often called the **Pontrjagin index** [4–6]. (Care should be taken not to confuse Q with the Pontrjagin numbers, which are defined as integrals of characteristic classes in $4n$ dimensions [2, 3, 7]).)

Using the chain rule, we can rewrite A in the form

$$A = \frac{1}{2}\epsilon_{abc}\phi^a d\phi^b \wedge d\phi^c, \tag{5.19}$$

which is now a two-form on the final sphere, rather than the initial square. Written in this form, it can be seen by comparison to standard formulas for solid angle that the integral Q is measuring $1/4\pi$ times the solid angle on S^2 filled by the image of the map ϕ. Since the solid angle of a sphere is 4π, it follows that Q is measuring the number of times the initial compactified sphere wraps around the final sphere.

It can be shown that Q is integer-valued, and it is a homotopy invariant: it is unchanged by smooth deviation of the map ϕ. In fact, recall from the last section that the Chern number of a two-dimensional manifold M classifies maps ϕ: $M \to S^2$. Although the definitions given here for the Chern number and Pontrjagin index look very different, it shouldn't be too hard to see that the two invariants are actually the same in the case where M is a sphere.

The Pontrjagin index as defined here characterizes the homotopy class of a map $\boldsymbol{\phi}$ from a two-sphere embedded in the three-dimensional Euclidean space $\mathbb{R}^3$ to a two-sphere in an internal space. This invariant is used to describe the topological charge of static t'Hooft–Polyakov solitons in $\mathbb{R}^3$. An analog exists in four-dimensional Euclidean space $\mathbb{R}^4$, which characterizes the homotopy group $\pi_3(S^3)$ of maps from a three-sphere in physical Euclidean space-time to a three-sphere in an internal space. This four-dimensional Q gives the topological charge of Yang–Mills instantons in gauge theory. For discussions of t'Hooft-Polyakov solitons and Yang–Mills instantons, the reader can refer to sections 3.4 and 4.2 of [6].

5.6 Hopf index

The last section discussed maps from two-spheres to two-spheres. Now suppose that $\boldsymbol{\phi}$ is a map from the *three*-sphere to the two-sphere, $\boldsymbol{\phi}$: $S^3 \to S^2$. A volume form can be defined on the two-sphere by

$$\Omega = \frac{1}{4\pi}\sin\theta \, d\theta \wedge d\phi, \tag{5.20}$$

normalized so that $\int_{S^2}\Omega = 1$. The preimage (or **pullback**) under ϕ of this two-form on S^3 can be shown to be exact, or in other words, it is the exterior derivative $d\omega$ of some one-form ω. Then the **Hopf invariant** or **Hopf index** is defined by

$$H(\phi) = \int_{S^3}\omega \wedge d\omega. \tag{5.21}$$

This is again an integer-valued homotopy invariant, unchanged under smooth deformations of ϕ.

As a specific example, consider the Hopf fibration (section 4.7). The two-sphere can be covered by two open sets, $S^2 = U_0 \cup U_1$, where U_0 and U_1 are the sets formed by the sphere with the South pole or the North pole removed, respectively. Each of these sets can be stereographically projected onto the plane with coordinates x and y. For example if $\{\phi^1, \phi^2, \phi^3\}$ are coordinates for the $\mathbb{R}^3$ in which the sphere is embedded, then the projection of U_0 is:

$$x = \frac{\phi^1}{1 + \phi^3}, \qquad y = \frac{\phi^2}{1 + \phi^3}. \tag{5.22}$$

Inverting the projection, we find:

$$\phi^1 = \frac{2x}{1 + r^2}, \qquad \phi^1 = \frac{2y}{1 + r^2}, \qquad \phi^2 = \frac{1 - r^2}{1 + r^2}, \tag{5.23}$$

where $r^2 \equiv x^2 + y^2$, and $(\phi^1)^2 + (\phi^2)^2 + (\phi^3)^2 = 1$. The volume form on the two-sphere can be written as

$$\Omega = \frac{1}{\pi}\frac{dx \wedge dy}{(1 + r^2)^2}. \tag{5.24}$$

Then if we parameterize S^3 with coordinates $\{x, y, \psi\}$ where $0 \leqslant \psi < 2\pi$, then we can choose a two-form of the form

$$\omega = \frac{i}{4\pi}\mathrm{Tr}(\sigma_3 s^{-1} ds) = \frac{1}{2\pi}\left(\frac{x\, dy - y\, dx}{1 + r^2} + d\psi\right), \tag{5.25}$$

where $s \in SU(2) \sim S^3$. It can then be readily checked that the Hopf bundle projection from S^3 to S^2 has Hopf index

$$H(\pi) = \int_{S^3} \omega \wedge d\omega = \frac{1}{2\pi}\int_{S^3} \frac{dx \wedge dy \wedge d\psi}{(1 + r^2)^2} = 1. \tag{5.26}$$

The inverse image $\boldsymbol{\phi}^{-1}(p)$ of a point p in S^2 is a loop in S^3. It can be shown [8] that the Hopf invariant is equal to the linking number of the pair of loops formed by the inverse images of any two distinct points $p_1 \neq p_2$ in S^2. The linking number of a pair of loops is defined in the next section.

5.7 Linking number and other invariants

A **knot** is topologically a circle (a closed loop) embedded in a three-dimensional space, but the parts of the curve may pass over and under each other in such a way that the curve cannot be continuously deformed into a planar circle without crossings (figure 5.7(a)). 'Continuously deformed', of course, means the knot can not be cut and then reconnected afterward. If the curve can be continuously deformed into a simple circle, then it is called a trivial knot or an **unknot** (figure 5.7(b)). Similarly, two or more knots may be tangled in such a way that they can not be continuously separated from each other; these sets of tangled-up knots are called **links** (figure 5.7(c)). We assume that the knots and links we consider are oriented: each

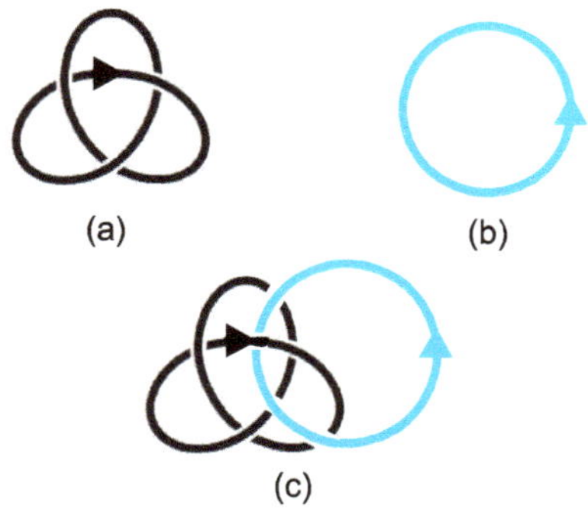

Figure 5.7. Examples of: (a) a knot (specifically a trefoil knot), (b) an unknot, and (c) a two-component link. The arrows give the orientation of each connected component.

component loop has an arrow attached, providing a direction in which to proceed around the loop.

If a knot is simple enough, then it may be easy to see by direct inspection whether or not it is an unknot. In general, however, the loop may be too complex for this to work. In this case, the strategy to distinguish whether a knot is an unknot, or whether two knots can be deformed into each other, is to compute topological invariants for the knots and then to compare the values. If the knots are equivalent, all of their invariants must be equal. If even one invariant differs between them, then they can not be deformed into each other. Many knot invariants have been constructed, such as the Jones polynomial and the Alexander polynomial. Similarly, invariants may be defined for links.

Here we mention just one type of invariant which can be defined for a single knot or for a link. (For other types of knot and link invariants, see references [9–12].) Consider two closed curves in $\mathbb{R}^3$. These can be described by the vectors $\boldsymbol{c}_1(s)$ and $\boldsymbol{c}_2(s)$, where $\boldsymbol{c}_{1,2}(s)$ are vectors giving the position of the curves at parameter value s, with $0 \leqslant s \leqslant 1$. Given parameterizations $\boldsymbol{c}_1(s_1)$ and $\boldsymbol{c}_2(s_2)$, then the **linking number** of the two curves is

$$L(\boldsymbol{c}_1, \boldsymbol{c}_2) = \frac{1}{4\pi} \int \frac{d\boldsymbol{c}_1}{ds_1} \cdot \left(\left(\frac{(\boldsymbol{c}_1 - \boldsymbol{c}_2)}{|\boldsymbol{c}_1 - \boldsymbol{c}_2|^3} \right) \times \frac{d\boldsymbol{c}_2}{ds_2} \right) ds_1\, ds_2 \tag{5.27}$$

$$= \frac{1}{4\pi} \int \left(\frac{(\boldsymbol{c}_1 - \boldsymbol{c}_2)}{|\boldsymbol{c}_1 - \boldsymbol{c}_2|^3} \right) \cdot (d\boldsymbol{c}_1 \times d\boldsymbol{c}_2). \tag{5.28}$$

If the two curves are not linked, then $L(c_1, c_2) = 0$; in this case, they can be continuously disentangled and separated from each other. Reversing the orientation of one connected component of the link will reverse the sign of the linking number.

Similarly, for a single knot we may define the **self-linking number**. This is done by taking the two curves in the previous definition to be copies of the same knot,

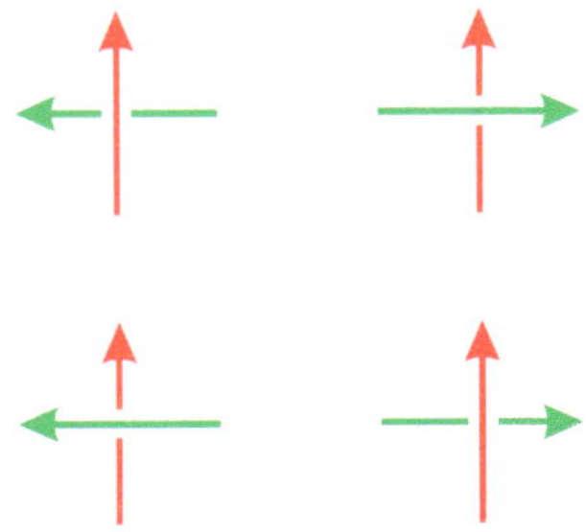

Figure 5.8. Positive crossings are those in which the arrow on top must rotate counterclockwise to align with the arrow on the bottom. The top arrow rotates clockwise for negative crossings.

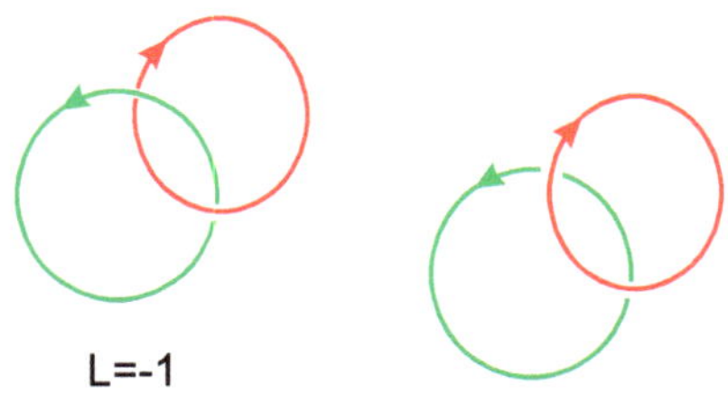

Figure 5.9. Examples of linking number. In (a), both crossings are negative, giving a linking number $L = 0$. In (b) there is one crossing of each sign, so that $L = 0$.

but slightly displaced from each other to avoid the denominator from being singular:

$$L(\boldsymbol{c}) \equiv L(\boldsymbol{c}, \boldsymbol{c}'), \tag{5.29}$$

where $\boldsymbol{c}'(s) = \boldsymbol{c}(s) + \boldsymbol{\epsilon}$ for some small displacement vector $\boldsymbol{\epsilon}$.

It can be shown that the linking number is an integer. In fact, it is given by the difference

$$L(\boldsymbol{c}_1, \boldsymbol{c}_2) = \frac{1}{2}(n_+ - n_-). \tag{5.30}$$

Here, n_+ is the number of positive crossings (those in which the oriented segment on the top of the crossing must rotate counterclockwise to align with the one under the crossing point), and n_- is the number of negative crossings (in which the segment on top rotates clockwise); see figure 5.8. This difference is invariant if the knot or link is rotated or deformed, since n_+ and n_- always change by the same amount, as the readers can readily convince themselves by experimenting with a few examples. Examples of linking number are shown in figure 5.9.

5.8 Atiyah–Singer index theorem

The Atiyah–Singer index theorem, first proven in 1963, has become famous because it provides a link between the topological properties of a manifold and the anaytical properties of differential operators defined on that manifold. The theorem has entered physics through several different applications, being used to study such things as magnetic monopoles, anomalies in quantum field theory, and Dirac operators (see [2, 13] and references therein). The flow of ideas has also run in the other direction, from physics to mathematics: proofs of the index theorem have been given using supersymmetry and the heat equation. Although it has not found widespread use in optics up to now, we discuss the Atiyah–Singer theorem briefly here. The description will be very cursory, since a more complete discussion would require defining characteristic classes, which would take us too far astray.

Consider a differential operator D on a manifold M. For simplicity, assume M is compact, oriented, and has no boundary. The differential operator could be something simple like the exterior derivative or Laplacian on M, or it could be a more complicated operator acting on sections of vector bundles over M, in which case, D could be viewed as a matrix-valued differential operator, with the matrix acting on the vector space that forms the fiber. To be specific, suppose D takes sections of some vector bundle E over M to section of a bundle F over M. If there is an inner product structure defined on the bundle then we can define an adjoint operator $D^\dagger$ that takes sections of F to sections of E: $\langle f|Dg\rangle = \langle D^\dagger f|g\rangle$, where f and g are sections of the two bundles. Then the **kernel** of D, written $\ker(D)$, is the space of solutions to the equation $Dg = 0$. Similarly, we can define the kernel of $D^\dagger$ as the space of solutions to $D^\dagger f = 0$. Then the **analytic index** is the difference in the dimensions of these two space:

$$\mathrm{Ind}_A(D, E) = \mathrm{Dim}\,\ker(D) - \mathrm{Dim}\,\ker(D^\dagger). \tag{5.31}$$

The **topological index** is an integral over the manifold of an expression built from the Chern character and Todd class, which we won't define here (see [2, 3, 7] for definitions), but it suffices to say that this index is a topological invariant.

The **Atiyah–Singer index theorem** then says that

$$\mathrm{Ind}_A(D, E) = \mathrm{Ind}_T(D, E). \tag{5.32}$$

The simplest example is when $D = d$ is the exterior derivative, acting on the bundle of differential forms over manifold M. Then for the two-dimensional case, the analytical index simply becomes the Euler characteristic, and the index theorem reduces back to the Gauss–Bonnet theorem. Other special cases include the Riemann–Roch theorem and the Hirzebruch signature theorem [2].

References

[1] Thouless D, Kohomoto M, Nightingale M and den Nijs M 1982 *Phys. Rev. Lett.* **49** 405
[2] Nakahara M 2003 *Geometry, Topology and Physics* 2nd edn (Boca Raton, FL: CRC Press)
[3] Steenrod N 1951 *The Topology of Fibre Bundles* (Princeton, NJ: Princeton University Press)
[4] Morandi G 1992 *The Role of Topology in Classical and Quantum Physics* (Berlin: Springer)
[5] Belavin A A and Polyakov A M 1975 *JETP Lett.* **22** 245
[6] Rajaraman R 1982 *Solitons and Instantons* (Amsterdam: Elsevier)

[7] Milnor J W and Stasheff J 1974 *Characteristic classes* (Princeton, NJ: Princeton University Press)
[8] Bott R and Tu L 1982 *Differential Forms in Algebraic Topology* (Berlin: Springer)
[9] Burde G, Zieschang H and Heusener M 2014 *Knots* (Berlin/Boston, MA: De Gruyter)
[10] Rolfsen D 1976 *Knots and Links* (Wilmington, DE: Publish or Perish)
[11] Kauffman L H 2001 *Knots and Physics* 3rd edn (Singapore: World Scientific)
[12] Baez J C and Muniain J P 1994 *Gauge Fields, Knots and Quantum Gravity* (Singapore: World Scientific)
[13] Nash C 1992 *Differential Topology and Quantum Field Theory* (London: Academic)

Chapter 6

Vortices and corkscrews: singular optics

6.1 Optical singularities

We are all familiar with vortices in fluid systems, from the whirlpool that forms around the drain in the kitchen sink to tornados that can devastate entire communities. A singular point is a point at which some quantity becomes undefined or a point at which a vector field vanishes. Singular points may be isolated, or may be part of a higher dimensional singularity, such as a one-dimensional vortex line. Vortices are singular points around which some vector field, for example the velocity field of a fluid or the phase gradient of a quantum field, circulates. These singular points and vortices are examples of **topological defects**: points, curves, or surfaces on which sudden, discontinuous changes occur in a system.

Much of the terminology in this area has been taken over from the study of topological defects in solids and of vortices in fluids, but singular points are common in optics as well, most often involving points where the phase or polarization of an optical field is undefined. At these points the amplitude of the electric field must vanish in order for the field itself to remain well-defined and single-valued. The field of **singular optics**, the study of optical fields in which such singularities occur, has been expanding rapidly as an active research area, and in this chapter a few topics from this area are discussed.

Singular optics can be divided into two parts: the study of singular points in ray optics and in wave optics. The chief example of a ray optical singularity is a **caustic** [1]. A caustic is a curve or surface on which many rays 'pile up' on top of each other, often along the envelope of a group of rays, to produce a region of very high intensity (see figure 6.1). These caustics are well-studied and provide examples of the mathematical structures known as **catastrophes** [2–4].

Our main concern here is with *wave* optical singularities. In particular, we will focus on optical waves with phase singularities and polarization singularities. Only relatively recently has it been recognized that the presence of singularities is in fact the rule in optical fields rather than the exception. Even a system as simple

doi:10.1088/978-0-7503-3471-6ch6

(a) (b)

Figure 6.1. (a) Caustics like the one here (the envelope along the upper left of the collection of rays) occur when rays 'pile up' on each other, forming curves or surfaces of high intensity. (b) An example of a caustic in geometric optics. The reflected rays from the side of the glass accumulate to a high intensity at the edges of the ray envelope, producing an extremely bright region.

as the Young two-slit experiment exhibits phase singularities [5]. In fact it turns out that the polarization and phase singularities discussed in the following sections are generic features of optical fields. (A property is called **generic** if it occurs with near certainty in a randomly prepared system, unless something is done to specifically prevent it. Generic properties of systems tend to be extremely stable.)

Although optical vortices and ray singularities had occasionally attracted the attention of luminaries such as Airy, Hamilton, Young and Stokes (as detailed by Berry [6], for example) since the early 19th century, the serious study of singular optics began only in the 1970s, especially with the publication of work by Nye and Berry [7]. More detailed reviews of singular optics can be found in [5, 8].

Singularities often arise as zeros of vector fields. We are all familiar with intensity zeros appearing in optics as a result of interference. For example, in the Young two-slit interference experiment we get a regular pattern of alternating bright and dark bands. At the center of each dark band there is a zero of intensity. By interfering more than two beams, arbitrarily complex patterns of bright and dark regions can be formed (figure 6.2). A common physical example is laser speckle, in which coherent light reflecting off a rough surface forms complex interference patterns filled with randomly placed singularities corresponding to intensity zeros. In three dimensions, these zeros can be at isolated points, they can form one-dimensional lines or filaments, or they may fill two-dimensional surfaces. This offers a wide range of features in which a variety of topological structures may potentially arise.

In the next section we will see a particular type of phase vortex which occurs in light beams with orbital angular momentum. We then briefly discuss vortices and topological defects more generally, before coming to polarization singularities.

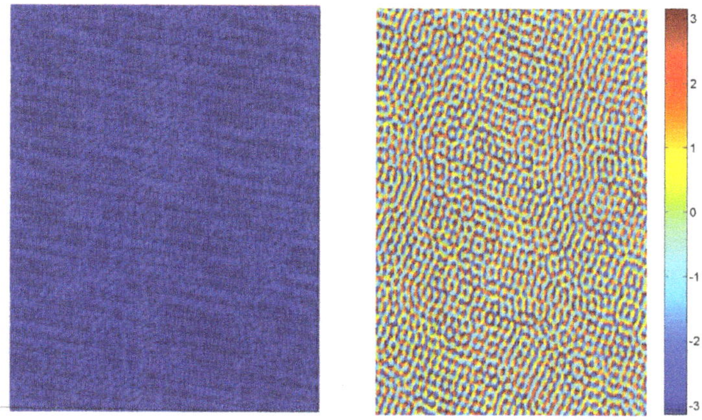

Figure 6.2. A collection of 50 randomly chosen plane waves produces a complex interference pattern. The intensity on a two dimensional plane is plotted on the left, and the corresponding phase structure on the right. Phase singularities often occur at intensity zeros.

6.2 Optical angular momentum

Angular momentum comes in two types in quantum mechanics: spin and orbital. Both play a role in optics. The role of spin angular momentum was recognized early on, since it leads to optical polarization, but orbital angular momentum (OAM) was largely ignored until the 1990s. Since then, it has been a hot topic as researchers have found a broad range of applications, including super-resolution microscopy, optical manipulation of nanoparticles, and quantum cryptography.

In classical mechanics, angular momentum is introduced by considering an object of mass m whose points all rotate about a fixed point P with constant speed v and orbital radius r. If the axis of rotation passes through the object itself, the object is said to be *rotating* or *spinning*; otherwise the object is said to be *orbiting* P. For a point mass, the angular momentum about the orbital axis is $\boldsymbol{L} = m\boldsymbol{v} \times \boldsymbol{r} = \boldsymbol{r} \times \boldsymbol{p}$, where $\boldsymbol{p}$ is the linear momentum and $\boldsymbol{r}$ is the vector pointing from the center of rotation to the point mass. For a non-point mass, the momentum *density* must be found and then integrated over the volume of the object.

Quantum mechanically, angular momentum is represented by a differential operator obtained in the position representation by making the usual replacement of momentum operators by spatial derivatives, $\boldsymbol{p} \rightarrow \hat{\boldsymbol{p}} = -i\hbar\nabla$. This leads to the operator

$$\hat{\boldsymbol{L}} = -i\hbar\, \hat{\boldsymbol{r}} \times \nabla. \tag{6.1}$$

For motion confined to a plane, this expression can be simplified to

$$\hat{L} = -i\hbar \frac{\partial}{\partial \phi}, \tag{6.2}$$

where ϕ is the angle about the rotation axis. Once a measurement axis is chosen (usually the z-axis), there are two commuting observables which can be simultaneously measured. One is the magnitude squared $\hat{L}^2 = \hat{L}_x^2 + \hat{L}_y^2 + \hat{L}_z^2$ of $\hat{\boldsymbol{L}}$, the other is its z-component $\hat{L}_z$. So if the wave function ψ is an angular momentum eigenstate, both of these operators have quantized eigenvalues:

$$\hat{L}^2\psi(\boldsymbol{r}) = l(l-1)\hbar^2\psi(\boldsymbol{r}), \qquad \hat{L}_z\psi(\boldsymbol{r}) = m_l\hbar\,\psi(\boldsymbol{r}), \tag{6.3}$$

where the discrete quantum numbers l and m have allowed values $L = 0, 1, 2, 3\ldots$ and $m_l = -l, \ldots, 0, \ldots, l$. The allowed wavefunctions are then labeled by the pair of integers (l, m): $\psi_{lm}(\boldsymbol{r})$.

The orbital angular momentum $\hat{L}$ results from motion of the particle through space; for example, the orbital motion of an electron around an atomic nucleus or of quarks around each other within a proton. However, even a particle without any orbital motion may have an *intrinsic* angular momentum built into the structure of its wavefunction. This **spin** angular momentum is of fixed magnitude, independent of the particle's motion, and is represented by an operator $\hat{\boldsymbol{S}}$. Its squared magnitude has a form similar to that of $\hat{L}^2$, with corresponding quantum number s:

$$\hat{S}^2\psi(\boldsymbol{r}) = s(s-1)\hbar^2\psi(\boldsymbol{r}). \tag{6.4}$$

But unlike l, the value of s is eternally fixed for a given type of particle: scalar field quanta like the Higgs boson have $s = 0$, while particles that mediate forces (photons, gluons, W and Z bosons) have $s = 1$; the sole exception is the graviton, which has spin 2. More generally, any boson (any particle not obeying the Pauli exclusion principle) has an integer value of spin, in units of $\hbar$. In contrast, matter particles such as electrons, protons, neutrinos, and quarks have $s = 1/2$; more generally, all fermions (particles that obey the exclusion principle) have s values that are odd multiples of 1/2. As in the angular momentum case, the component of $\hat{\boldsymbol{S}}$ along any axis will be quantized:

$$\hat{S}_z\,\psi(\boldsymbol{r}) = m_s\hbar\,\psi(\boldsymbol{r}), \tag{6.5}$$

where $m_s = -s, \ldots, 0, \ldots s$. The total angular momentum of a quantum state is then the sum of the spin and angular momenta:

$$\hat{\boldsymbol{J}} = \hat{\boldsymbol{L}} + \hat{\boldsymbol{S}}. \tag{6.6}$$

The various components of $\hat{\boldsymbol{S}}$ are not mutually commuting and so cannot be simultaneously measured. The same is true for the components of $\hat{\boldsymbol{L}}$. However, the components of $\hat{\boldsymbol{L}}$ do commute with the components of $\hat{\boldsymbol{S}}$, so that spin and orbital angular momentum measurements do not affect each other.

The **helicity** of a particle is the component of total angular momentum along the direction of the particle's motion,

$$\hat{h} = \frac{\hat{\boldsymbol{J}}\cdot\hat{\boldsymbol{p}}}{|p|} = \frac{\hat{\boldsymbol{S}}\cdot\hat{\boldsymbol{p}}}{|p|}. \tag{6.7}$$

The last equality occurs because the orbital angular momentum $\hat{\boldsymbol{L}} = \hat{\boldsymbol{r}} \times \hat{\boldsymbol{p}}$ is orthogonal to $\boldsymbol{p}$, so the helicity is determined entirely by the spin. Photons in a beam of light should have possible helicity values $m_s = -1, 0, +1$, in units of $\hbar$. However, gauge invariance [9–11] forces the $m_s = 0$ component to vanish for propagating modes of massless particles like photons. (This is not necessarily true for the *virtual* photos that form the Coulomb field or the evanescent waves that occur at boundaries between optical materials, but we consider only propagating optical fields here.) So despite being a boson, light has two possible spin states, like a spin-1/2 fermion. Because of this, photon polarization and electron spin can each carry one qubit of information in quantum information processing applications, and both particles can be treated with the same two-state formalism. Polarization is discussed in more detail in chapter 2, and will be important in the discussion of geometric phases in chapter 9. For now, we focus on orbital angular momentum.

We will take $\hat{L}_z$ to be the component of orbital angular momentum about the propagation axis, and to conform with the more common notation in optics we will now denote the L_z eigenvalue that we had previously referred to as m_l simply by m: $m_l \to m$. Since it is hard to see how light traveling along the z-axis to be in any sense rotating about that axis, the detailed study of orbital angular momentum (OAM) in optics got off to a much later start than the study of polarization, only taking off in the 1990s, following the publication of [12]. The OAM in this case arises as a result of having a wavefront with nontrivial spatial structure in the transverse direction. One can view the angular momentum as arising from the rotation of this wavefront, and therefore of the resulting rotation of the Poynting vector. There are now many reviews of both the physics of optical OAM and its applications, including [13–17].

Let us consider the simplest possible case. Multiply an approximate plane wave by an azimuthally dependent phase shift of the form $e^{im\phi}$, where ϕ is the angle about the propagation axis (figure 6.3). The angular momentum operator $\hat{L}_z = -i\partial/\partial\phi$ clearly has eigenvalue $L_z = m\hbar$. The wavefunction must be single-valued, so invariance under the transformation $\phi \to \phi + 2\pi$ forces the OAM quantum number m to be quantized: $e^{im\phi} = e^{im(\phi+2\pi)}$ can only be true if m is an integer. Since the azimuthally increasing phase factor has the effect of tilting the wavefronts by an increasing amount as one circles the axis, the wavefronts end up with a corkscrew shape, as in figure 6.4. The Poynting vector $\boldsymbol{S}$ orthogonal to the wavefront therefore also rotates, leading to nonzero OAM.

m is an example of a topological charge; more specifically, it is a winding number. The phase can be viewed as a map $\Phi: S^1 \to S^1, \quad \Phi(\phi) = e^{im\phi}$, where the first circle is spanned by the angle ϕ around the propagation axis and the second is the unit circle traced out by the exponential $e^{im\phi}$ in the complex plane. The winding number counts the number of times the first circle wraps around the second.

Each direction at which one can move radially outward from the axis corresponds to a different value of the phase. On the axis, all of these directions collapse to the same point, so the phase is clearly not well defined there. In order for the optical field as a whole to remain well-defined everywhere, it is necessary that the amplitude vanish at the origin. The value of phase at that point then becomes irrelevant.

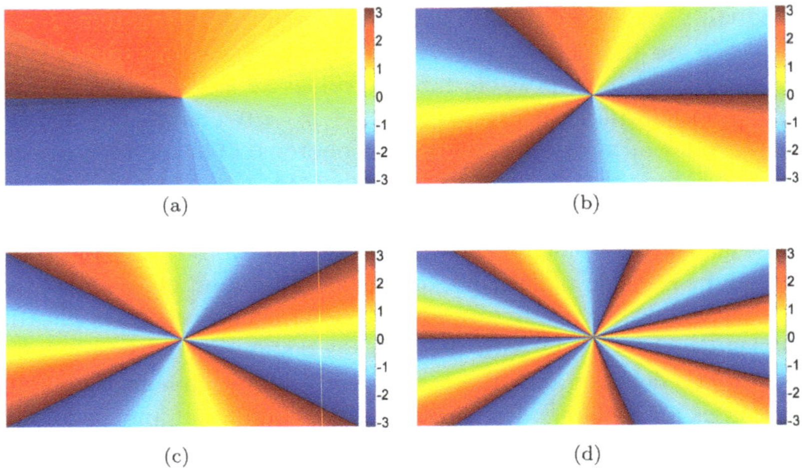

Figure 6.3. The phase of OAM modes in the transverse plane. The origin of the plot is at the z-axis. A state with winding number m has a phase that varies by $2\pi m$ as the azimuthal angle completes one circuit from 0 to 2π. (a)–(d) respectively show the phase for the cases $m = 1, 3, 4, 7$.

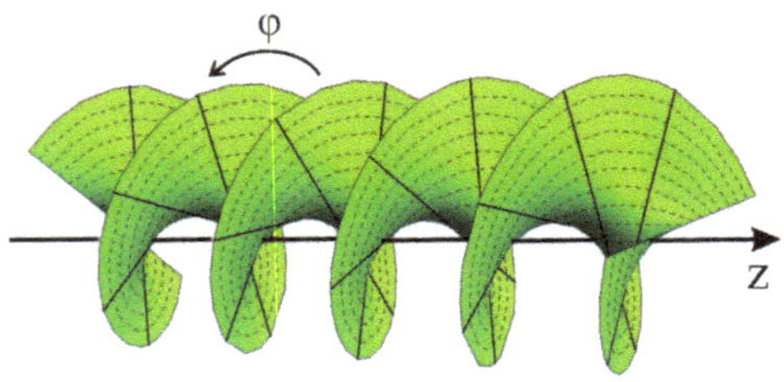

Figure 6.4. Twisted light: an optical wavefront with nonzero orbital angular momentum. Surfaces of constant phase are corkscrew-shaped. The Poynting vector, $\boldsymbol{S}$, must be everywhere perpendicular to the wavefront, so that it rotates as the corkscrew propagates along the z-axis.

This singular point effectively forms a hole around which the azimuthal angle circles. The OAM-based topological charge then labels the distinct homotopy classes on this punctured configuration space, with each distinct value of m labeling a homotopy class of maps $e^{(im\phi)}$ circling the puncture m times.

Many different optical beam modes can carry OAM, including higher-order Bessel or Hermite–Gauss modes [16]. Here we only discuss the Laguerre–Gauss (LG) modes. Each mode is characterized by two integers, the OAM quantum number m and the radial quantum number p which counts the number of nodes in the radial direction. The explicit form for the beam amplitude is given by [18]

$$E_{mp}(r, z, \phi) = \frac{E_0}{w(z)}\left(\frac{\sqrt{2}\,r}{w(z)}\right)^{||} e^{-r^2/w^2(r)} L_p^{|m|}\left(\frac{2r^2}{w^2(r)}\right) \times e^{-ikr^2z/\left(2(z^2+z_0^2)\right)} e^{-i\phi m + i(2p+|m|+1)\arctan(z/z_0)}, \tag{6.8}$$

where E_0 is a constant and $w(z) = w_0\sqrt{1 + z/z_0}$ is the beam radius at distance z. $L_p^\alpha(x)$ are the associated Laguerre polynomials [19], $z_0 = \pi w_0^2/\lambda$ is the Rayleigh range, and the arctangent term is the Gouy phase. Here we work in cylindrical polar coordinates: r is the distance from the axis in the transverse plane, z is the distance along the propagation axis, and ϕ is the azimuthal angle about the axis. The lowest order LG mode, $m = p = 0$ is the standard Gaussian beam. Examples of the intensity profiles of several values of m and p are plotted in figures 6.5 and 6.6.

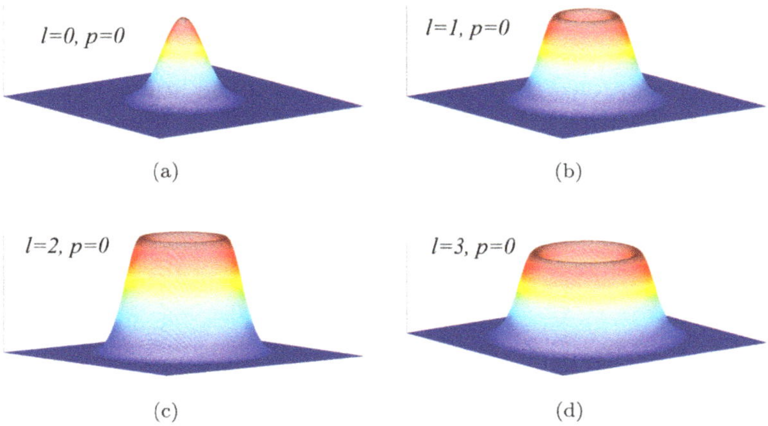

Figure 6.5. Intensity profiles in the transverse plane for several $p = 0$ *LG* modes with different values of topological charge m. The $m = 0$ case is the standard Gaussian beam, with an intensity maximum on the axis. In contrast, the intensity vanishes on the axis for the $m \neq 0$ cases, due to the presence of a phase singularity there.

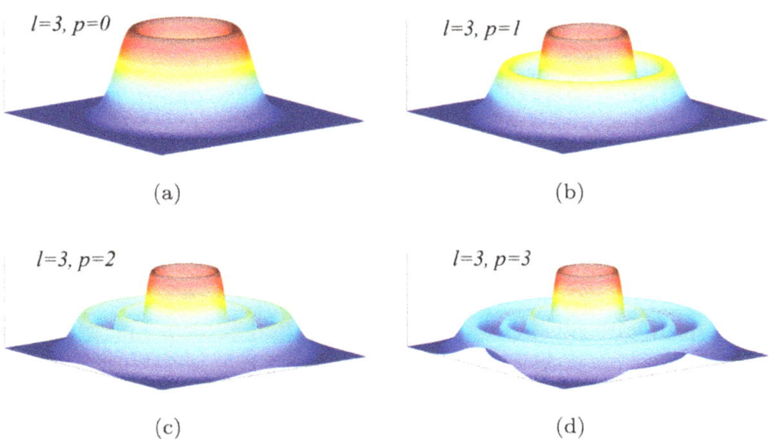

Figure 6.6. Intensity profile in the transverse plane for several $m = 3$ *LG* modes with different p values. Increasing p increases the number of radial intensity nodes, but has no effect on the azimuthal distribution of phase or intensity.

The point on the axis in the transverse plane is a **phase singularity**, a point at which the phase is undefined, forcing the amplitude to vanish. This singularity is an example of a **vortex**. Vortices will be discussed in more generality in the next section, but in the two-dimensional transverse plane the vortex is a single point. Adding in the longitudinal direction, this extends to a vortex line or vortex curve in the full three-dimensional space. In the context of the Laguerre–Gauss beam, the vortex line is simply the propagation axis, but we will see in chapter 7 that more general vortex lines can be more complicated, even forming knots and links.

Our main interest here is the behavior of the beam as one winds about the singular dark spot at the axis. This behavior is determined by the topological charge m. But it should be noticed that p also has topological significance. For $p \neq 0$, u_{lp} has p dark nodal rings circling the axis. There will be curves that can become trapped between the rings, adding additional topological structure and making the homotopy classes more complicated. The physical meaning of the radial quantum number is not as apparent as that of m, but it has interesting properties in its own right. For example, it has been shown that a set of raising and lowering operators can be defined that carry the system between different p states [20–22].

Optical OAM beams have been proposed for many types of applications in both classical and quantum communication schemes and in quantum cryptography [23–28]. Photonic OAM has a major advantage over using photon polarization. The latter has only two values and so can carry only a single bit of classical information or a single qubit of quantum information. In contrast, if a photon can be produced with OAM values ranging from $-M$ to $+M$, then up to $\log_2(2M+1)$ bits or qubits can be carried by a single photon and in principle M can be very large.

But there are practical difficulties with sending OAM states over long distances. For example, designing multiple-mode optical fibers that will carry a large range of OAM values is not easy, although progress is being made. The other possibility is to transmit the vortices through the open air; but this runs into problems with atmospheric turbulence. If a mode with topological charge m is sent through a turbulent atmosphere, it tends to break up into multiple vortices of lower OAM. There are various processes by which that can occur, such as the creation of a pair of vortices of equal and opposite OAM, $\pm m$, or the merging of two vortices (topological charges m_1 and m_2) into a single vortex of value $m_1 + m_2$. In each case, the total OAM value is conserved. However, the vortices tend to wander randomly due to the turbulence and to move apart due to ordinary diffractive spread, so that by the time the light reaches the detector on the receiving end of the communication link, many of the newly formed vortices will miss the detector [29]. The net result is that over multiple trials the experimenter will see the total m value fluctuating randomly about the originally transmitted values. The problem becomes worse as the transmission distance or the degree of turbulence increases, or as the detector size decreases. Nevertheless, a number of experiments in recent years have managed to send OAM-based signals over long-range free-space links [30–35].

The effect of the turbulence is to distort the lines of constant phase, as seen in figure 6.7. In the unperturbed case, these would be straight lines emanating radially outward from the vortex. The turbulence introduces random phase fluctuations due

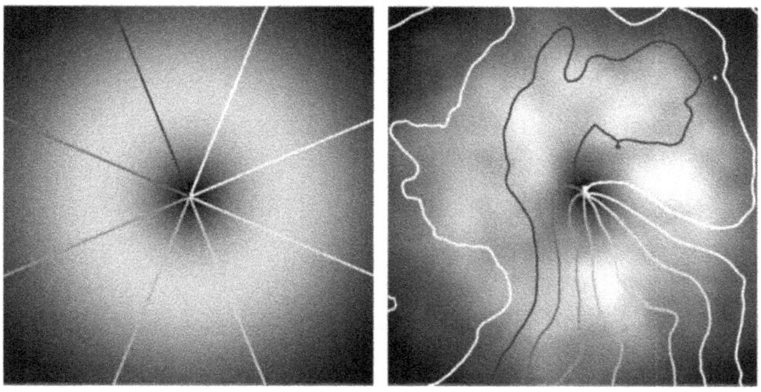

Figure 6.7. (a) In the absence of any perturbations, the phase increases azimuthally around the OAM vortex, so that the curves of constant phase are straight lines stretching radially outward from the singular point on the propagation axis. (b) Random phase fluctuations in the atmosphere introduce random deformations of the phase curves, which can cause the vortex to wander away from the axis. (Figures reproduced with permission from [47], copyright 2005 American Physical Society.)

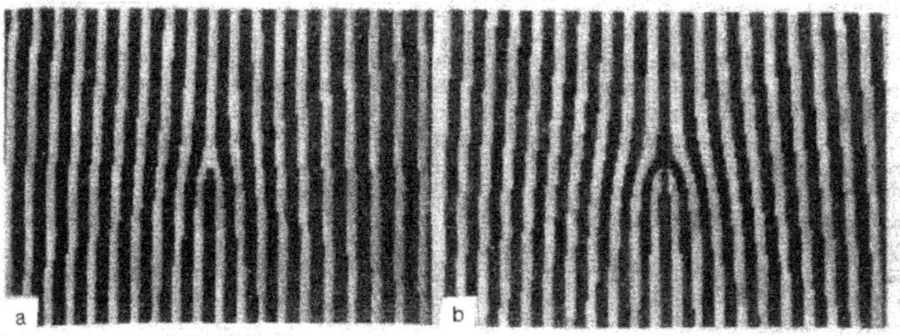

Figure 6.8. Examples of pitchfork diffraction gratings for changing the OAM of light. The shapes of these patterns represent edge dislocations. (Figures reproduced with permission from [48], copyright 1991 American Institute of Physics.)

to thermally induced refractive index changes at each point. The result is that the phase lines become randomly snaking curves, and the vortex, which lies at the intersection of all of these curves, will therefore exhibit a Brownian motion-like random walk. These are continuous deformations of the phase curves, and so they leave the topological charge unchanged. However, if the disturbances of the phase curves become strong enough, discontinuities can occur as the lines cross; this leads to vortex–antivortex pair creation. The two oppositely charged vortices then tend to wander away from each other, causing the conserved OAM to diffuse away from its initial point in the transverse plane.

In addition to communication and cryptography, applications of OAM in imaging, microscopy, and object identification also abound [36–46]. These applications include both classical and quantum entanglement-based methods, and some can provide super-resolving imaging that exceeds the classical Abbé and Rayleigh resolution limits.

Beams with nonzero OAM can be generated several ways, for example by using spatial light modulators or by sending light through a cylindrical piece of glass whose thickness increases around the azimuthal direction. The earliest papers on OAM generally used diffraction gratings of the form shown in figure 6.8.

6.3 Vortices and dislocations

In the last section, the Laguerre–Gauss optical fields formed vortices, with the phase varying linearly as one circulates around the singularity. Vortices, in which some physical variable circulates around a singular point, occur in many physical systems, from Bose–Einstein condensates and superconductors [49, 50], to Higgs field vortices in particle physics and cosmology [10, 51–53]. We are all familiar with vortices from fluid mechanics, ranging from whirlpools in a bathtub on a small scale to tornados, hurricanes, and Jupiter's great red spot on much larger scales. Trails of vortices (the **von Karmen street**) tend to appear in turbulent fluids and can be found along the edges of airplane wings, or in rising cigar smoke.

Vortices are characterized by the fact that some vector field $\boldsymbol{v}$ has a nonzero curl, called the **vorticity**,

$$\boldsymbol{\omega} = \nabla \times \boldsymbol{v}, \tag{6.9}$$

and a nonzero **circulation**

$$\Gamma = \oint_C \boldsymbol{v} \cdot d\boldsymbol{l} = \int_S \boldsymbol{\omega} \cdot d\boldsymbol{A}, \tag{6.10}$$

where $\mathcal{C}$ is any curve enclosing the singular vortex point, and $\mathcal{S}$ is any simply connected surface bounded by $\mathcal{C}$. The last equality follows by means of Stokes' theorem. These formulas should look familiar: they are directly analogous to the Gauss and Ampère law formulas, and if $\boldsymbol{v}$ is the phase gradient then the circulation is proportional to the topological charge m of the previous section.

As mentioned in the previous section, vortices form points in two dimensions, but extend into lines or curves in three dimensions. In order to be general, consider a complex scalar field $u(\boldsymbol{r}) = A(\boldsymbol{r})e^{i\psi(\boldsymbol{r})}$, which could represent, for example, a quantum wavefunction or the amplitude of one polarization component of an electric field. u is a solution to the Helmholtz equation or some other wave equation. By taking the gradient of u a vector field is constructed, and topological invariants arise in connection with the singularities in the vector field. The topological charge m of the field and the vector field index n (section 5.3) of its gradient are defined by

$$m = \frac{1}{2\pi}\oint_C \nabla\psi(\boldsymbol{r}) \cdot d\boldsymbol{r} = \mathcal{I}_P(\nabla\psi) \tag{6.11}$$

$$n = \frac{1}{2\pi}\oint_C \nabla\theta(\boldsymbol{r}) \cdot d\boldsymbol{r} = \mathcal{I}_P(\nabla\theta), \tag{6.12}$$

where $\mathcal{C}$ encloses the singularity and θ is the angle at which $\nabla\psi$ points relative to the x-axis. These are the indices $\mathcal{I}_P$ of the respective gradient field singularities. Notice that, aside from the normalization, they can be viewed as the circulations of the gradients. Stationary points of the phase, points at which $\nabla\psi$ vanish, can come in a number of types, including vortices, sinks, sources, and saddles (figure 6.9), which can be distinguished from each other by the values of winding number and index. The analyticity of the underlying differential equation satisfied by ψ will guarantee

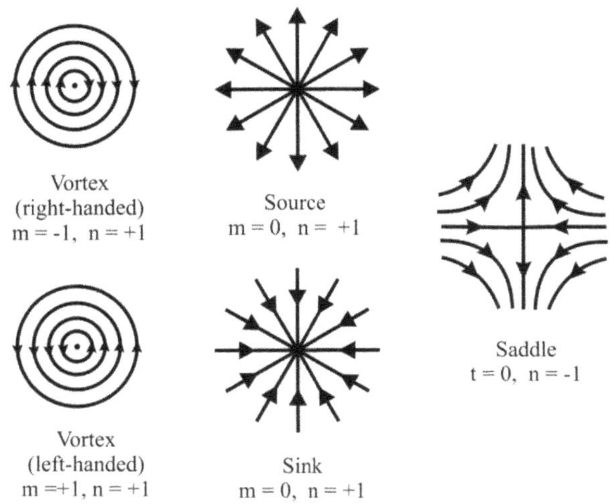

Figure 6.9. Examples of simple singularities in the gradient of a complex scalar field. As the a complete loop around the singularity is traversed counterclockwise, m counts the number of times the phase of the scalar field wraps around the unit circle, while n counts the number of times its gradient field rotates around the circle.

that these stationary points can only be created and annihilated in pairs, as we saw for the OAM vortices in the last section.

More generally, both topological charge and the sum of indices of all the singularities should be conserved, placing strong restrictions on the topological reactions that can occur. In particular, to conserve the total index some additional singular points must be created or annihilated along with the vortices. Some allowed interactions include:

- Two vortices, one right-handed ($m = +1$, $n = +1$) and one left-handed ($m = -1$, $n = +1$), annihilating along with two saddle points ($m = 0$, $n = -1$).
- A sink and a saddle (both $m = 0$, $n = 1$ each) annihilating to create a pair of oppositely handed vortices.

Phase singularities are also known as **screw dislocations**, where the terminology is borrowed from the study of topological defects in crystals. In a screw dislocation the wavefronts circulate around the axis (figure 6.10(b)). Another possibility is an **edge dislocation**, in which the number of wavefronts (or the number of lattice planes in the crystal case) discontinuously changes (figure 6.10(a)). The diffraction gratings in figure 6.8 show examples of edge dislocations. Edge dislocations also lead to circulating vortices, but their structure is more complicated and we won't discuss them further here; see [5] for a detailed discussion.

6.4 Polarization singularities

Light leaving the Sun is largely unpolarized, but scattering in the atmosphere can lead to a complex polarization pattern in the sky. This polarization of natural light may have had more than academic interest for sailors in previous centuries: lacking magnetic compasses, there is some evidence that Vikings made use of birefringent

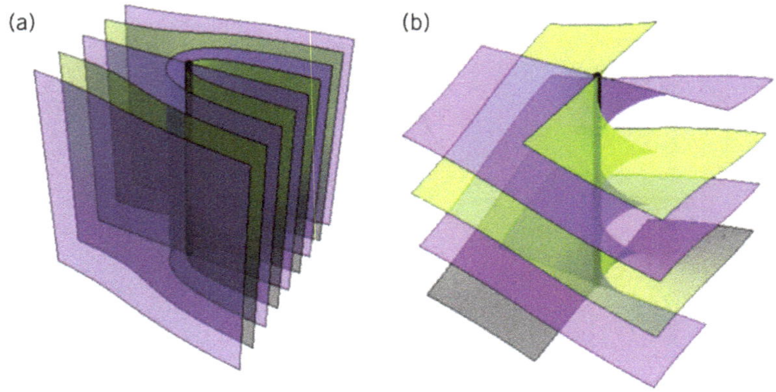

Figure 6.10. Typical wave dislocations: (a) edge dislocation and (b) screw dislocation. Mixed edge-screw dislocations are also possible [5]. (Figures reproduced with permission from [8], copyright 2009 Elsevier.)

crystals to measure this polarization pattern and that they used it to navigate when the Sun or stars were hidden by clouds [54].

In addition to the phase singularities of the previous sections, **polarization singularities** are common. These are points, lines, or surfaces on which the polarization direction becomes undefined or at which there is a transition between different types of polarization (linear versus elliptical). Like phase singularities, polarization singularities are surprisingly common in optics, even occurring in natural sunlight. As far back as the 1840s scientists such as Arago, Babinet, and Brewster noted that there are directions in the sky at which the degree of polarization vanishes [8]; these were called **neutral points**. Neutral points are examples of polarization singularities. We now know that polarization singularities, like phase singularities, are actually generic features of optical fields and are stable under smooth variations of the field.

Neutral points are singularities of the **polarization azimuth**, θ, defined as the angle at which the amplitude of the electric field oscillation is maximum. There are several ways in which θ can become undefined at a given point: the light can become unpolarized, the field amplitude can become zero, or the polarization can become circular so that no direction can be single out as the direction of a maximum.

It should be noted that the polarization direction does not define a vector, since polarization along unit vectors $\hat{n}$ and $-\hat{n}$ are equivalent. Rather, the polarization direction is an example of a **director**, an undirected line in space. In mathematical terms, vectors live on the Euclidean spaces $\mathbb{R}^n$, while directors live on the **real projective spaces** $\mathbb{RP}^n$ defined by identifying opposite directions. (See, for example, [55–58] for more detail on real and complex projective spaces.) Directors come up in a number of optical and physical contexts, for example in descriptions of liquid crystals [59]. Because opposite directions are equivalent for a director, director fields can have singularities with half-integer values of index, unlike vector field singularities which must always have integer index.

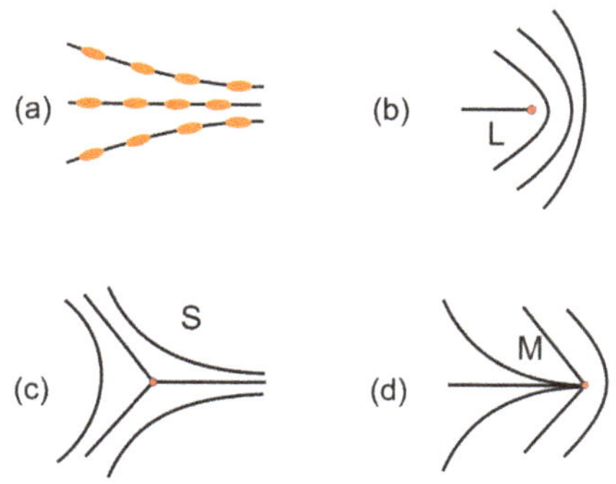

Figure 6.11. (a) The major axes of the polarization ellipses (the orange ellipses) can be strung together to form streamlines. Streamline flows of three types form around *C* points: (b) Lemon, (c) star, and (d) monstar. The gradients fields have respective indices of +1/2, −1/2, +1/2.

Ignoring the case of unpolarized light, consider an elliptically polarized light field. Elliptical polarization is in a sense the most general type of polarization, since other polarization types can be considered as degenerate cases of it: circular polarization occurs when the two axes of the polarization ellipse become equal in length, while linear polarization occurs when the ellipse collapses to a line segment with one of the axes shrinking to zero length. These two types of degenerate elliptical states correspond to the two common types of singularities in elliptically polarized light. A ***C* point** in two dimensions (or ***C* line** in 3D) is a point where the polarization ellipse becomes circular so that the director and the azimuth both become undefined. An ***L* line** in 2D (***L* surfaces** in 3D) occurs when the polarization becomes linear; in this case the director and the azimuth both remain well-defined, but the helicity (left- or right-handedness of the field) becomes indeterminate. ***L*** lines often occur as boundaries between regions of elliptical polarizations with opposite helicity.

The polarization azimuth can be viewed as the phase of a complex field defined in terms of the Stokes parameters section 2.4 by

$$\sigma = S_1 + iS_2 = |\sigma| e^{i\theta/2}. \tag{6.13}$$

C points occur when $S_1 = S_2 = 0$, and from the form of the phase factor in σ it is clear that they will have half-integer index. In contrast, *L* lines occur when $S_3 = 0$ (no circular polarization) and they have integer index of values ±1.

A line segment can be drawn along the major axis of the polarization ellipse at each point, and the segments at each point can be joined together to form continuous streamlines. In the vicinity of a *C* point, the streamlines can have three types of behavior, shown in figure 6.11; these three types are called a **lemon** (index +1/2), **star** (index −1/2), and a **monstar** (index +1/2). At a lemon, one streamline terminates, at a star three of them do. At a monstar, an infinite number of streamlines end, but with only three of them being straight lines. The monstar gets its name because it has the same number of straight terminating lines as the star but the same index as a lemon, and it is in some sense a transitional form between the other two. Extensive experiments have been done mapping *C* and *L* lines of optical systems and studying their physical and statistical properties; some examples involving polarization singularities in optical speckle fields include [60–62].

Although C- and L-points are the generic singularities that typically appear in polarized fields, nongeneric singularities can easily be produced by making appropriate arrangements. Maybe the most useful examples are **vector beams**, in which the polarization of an optical beam (a laser beam, for example) is spatially dependent and forms a well-defined pattern as one circulates around the beam axis. Such vector polarization fields will have singularities at the origin. The simplest examples are the radially polarized and azimuthally polarized examples in figure 6.12. Such beams are easily constructed from superpositions of orthogonally polarized Hermite–Gauss, Bessel–Gauss beams, or Laguerre–Gauss beams, all of which naturally arise in lasers or can be easily produced by standard manipulations of laser beams [16, 63].

Vector beams have found a number of applications over the past 15 years, in areas such as optical trapping, manipulation of nanoparticles, laser machining, and high resolution imaging. For a review of the theory and applications, see [64].

Azimuthally and radially polarized beams may be the simplest examples, but more exotic vector beams are also possible. For example, so-called **Poincaré beams** [65, 66] have been produced, in which every transverse cross-section of the beam spans the entire range of possible polarization states, covering the entire surface of the Poincaré sphere. In fact it is possible to go further, filling the entire range of partially polarized states in the *interior* of the sphere [66]. Such beams can be produced by passing circularly polarized light through a material with appropriate spatially varying birefringence, such as can be made to occur in a pane of glass placed under stress, and may be useful for transmitting optical information through turbulence [67, 68].

A simple example [5] of a Poincaré beam is given by

$$E(\boldsymbol{r}, \gamma) = \cos\gamma\ \hat{e}_1\ u_{00}(\boldsymbol{r}) + \sin\gamma\ \hat{e}_2\ u_{01}(\boldsymbol{r}), \tag{6.14}$$

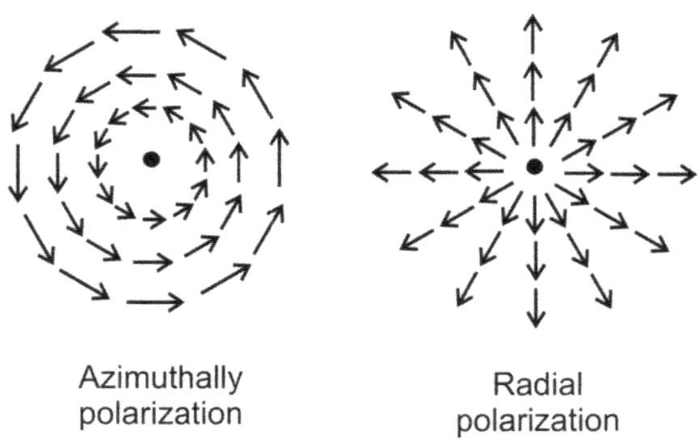

Figure 6.12. Cross sections of two vector beams, beams with spatially inhomogeneous polarization patterns. These beams can be formed from superpositions of Hermite–Gauss modes of lasers. The singularity at the center of each of these vector fields has topological index of +1.

where γ is a real paramater, u_{ij} are Laguerre–Gauss modes, and $\hat{e}_1$, $\hat{e}_2$ are a pair of orthogonal unit vectors perpendicular to the progation (z) axis. The components of the normalized Stokes vector, $s_i = S_i/|\boldsymbol{S}|$ are of the form

$$s_1 = \frac{2\bar{\rho}\cos(\phi - \zeta(z))}{1 + \bar{\rho}^2}, \qquad s_2 = \frac{2\bar{\rho}\sin(\phi - \zeta(z))}{1 + \bar{\rho}^2}, \qquad s_3 = \frac{1 - \bar{\rho}^2}{1 + \bar{\rho}^2}, \tag{6.15}$$

where the dimensionless parameter $\bar{\rho}$ is given in terms of the beam waist w and the radial variable r in the transverse plane as $\bar{\rho} = \sqrt{2}\tan \gamma r/w$. Inserting explicit forms for the Laguerre–Gauss modes, it is readily shown [5] that this field has the form

$$E(\boldsymbol{r}, \gamma) = C(\boldsymbol{r}, \gamma)[\hat{e}_1 + \bar{\rho}e^{i(\phi-\zeta(z))}\hat{e}_2], \tag{6.16}$$

where the prefactor $C(z, \phi, r, \gamma)$ has no effect on the polarization. Because of the form of r and ϕ dependence in the bracketed portion, every possible polarization state is guaranteed to occur somewhere in each transverse plane of fixed z. Suppose that the unit vectors $\hat{e}_1$, $\hat{e}_2$ are chosen to represent the directions of left- and right-handed polarizations, respectively, on the Poincaré sphere. Then, as the radial variable r ranges from 0 to ∞, the polarization evolves from left-circular through linear to right-circular. Different choices of $\hat{e}_1$ and $\hat{e}_2$ will lead to other polarization patterns in the plane.

6.5 Optical Möbius strips

Up to this point the surfaces have appeared in our discussions have been mainly oriented surfaces like spheres and tori. However, non-orientable surfaces can also occur in optics. We briefly mention one example [69–74] here.

Consider a field of elliptically polarized light. On C-lines, the major and minor axes of the polarization ellipse become equal, so that the elliptical polarization becomes circular. Suppose that a closed curve $\mathcal{C}$ encircles this C-line. As one traces out one complete circuit of this curve, the major and minor axes of the polarization ellipse both twist around the curve, rotating through some angle Ω. The polarization axes have to be single valued, and a polarization direction of $\hat{\boldsymbol{n}}$ is physically equivalent to one pointing along $-\hat{\boldsymbol{n}}$. So when one full loop is completed, Ω must equal either an integer or half-integer multiple of 2π. So defining the **twist index**

$$\tau = \frac{\Omega}{2\pi}, \tag{6.17}$$

τ is a conserved topological quantum number taking on only integer or half-integer values.

As $\mathcal{C}$ is traversed, a unit vector pointing along one of the polarization ellipse axes sweeps out a ribbon (figure 6.13), with τ counting out the number of π twists of the ribbon before it reconnects with itself. If $\tau = 1/2$, the vector rotates by π and the result is a non-orientable Möbius strip. For $\tau = 1$, the strip gets a full 2π twist, leaving an orientable ribbon. More generally, the result is orientable when τ is integer and non-orientable when τ is a half (odd) integer. In principle, any integer

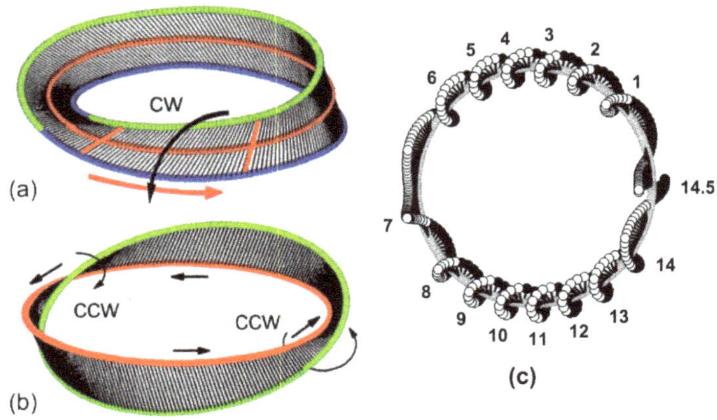

Figure 6.13. The polarization ellipse axes trace out ribbons with an integer number of π twists. The images on the left correspond to (a) $\tau = 1/2$ and (b) $\tau = 1$; the image on the right (c) has 29 π twists. (Figures (a) and (b) reproduced with permission from [74], copyright 2014 Optical Society of America, and (c) from [70], copyright 2014 Optical Society of America.)

number of π twists can be inserted, with positive and negative τ counting counter-clockwise and clockwise twists around the curve, respectively.

A method of producing such twisted polarization configurations has been proposed using superpositions of circularly polarized Laguerre–Gauss beams [70], and such optical Möbius strips have been experimentally demonstrated using liquid crystal q-plates [75].

References

[1] Nye J 1999 *Natural Focusing and Fine Structure of Light: Caustics and Wave Dislocations* (Boca Raton, FL: CRC Press)

[2] Arnol'd V I 1984 *Catastrophe Theory* (Berlin: Springer)

[3] Saunders P T 1980 *An Introduction to Catastrophe Theory* (Cambridge: Cambridge University Press)

[4] Gilmore R 1993 *Catastrophe Theory for Scientists and Engineers* (Mineola, NY: Dover)

[5] Gbur G J 2017 *Singular Optics* (Boca Raton, FL: CRC Press)

[6] Berry M 2000 *Nature* **403** 21

[7] Nye J F and Berry M V 1974 *Proc. Roy. Soc. Lond.* A **336** 165

[8] Dennis M R, O'Holleran K and Padgett M J 2009 *Prog. Opt.* **53** 293

[9] Zee A 2010 *Quantum Field Theory in a Nutshell* 2nd edn (Princeton, NJ: Princeton University Press)

[10] Ryder L H 1996 *Quantum Field Theory* (Cambridge: Cambridge University Press)

[11] Itzykson C and Zuber J-B 1980 *Quantum Field Theory* (New York: McGraw-Hill) reprinted by Dover, Mineola, NY, 2005

[12] Allen L, Beijersbergen M W, Spreeuw R J C and Woerdman J P 1992 *Phys. Rev.* A **45** 8185

[13] Yao A M and Padgett M J 2011 *Adv. in Opt. and Phot.* **3** 161

[14] Torres J P and Torner L (ed) 2011 *Twisted Photons: Applications of Light with Orbital Angular Momentum* (Hoboken, NJ: Wiley)

[15] Franke-Arnold S, Allen L and Padgett M 2008 *Laser Photon. Rev.* **2** 299
[16] Simon D S 2020 *A Guided Tour of Light Beams: From Lasers to Optical Knots* 2nd edn (Bristol: IOP Publishing)
[17] Simon D S, Jaeger G and Sergienko A V 2017 *Quantum Metrology, Imaging, and Communication* (Berlin: Springer)
[18] Allen L, Padgett M and Babiker A V 1999 *Prog. Opt.* **39** 291
[19] Arfken G, Weber H and Harris F E 2012 *Mathematical Methods for Physicists: A Comprehensive Guide* 7th edn (London: Academic)
[20] Karimi E and Santamoto E 2012 *Opt. Lett.* **37** 2484
[21] Karimi E, Boyd R W, de Guise H, Řeháček J, Hradil Z, Aiello A, Leuchs G and Sánchez-Soto L L 2014 *Phys. Rev.* A **89** 063813
[22] Plick W N, Lapkiewicz R, Ramelow S and Zeilinger A 2013 arXiv:1306.6517 [quant-ph]
[23] Vaziri A, Weihs G and Zeilinger A 2002 *Phys. Rev. Lett.* **89** 240401
[24] Gröblacher S, Jen New ein T, Vaziri A, Weihs G and Zeilinger A 2006 *New J. Phys* **8** 75
[25] Simon D S, Lawrence N, Trevino J, dal Negro L and Sergienko A V 2013 *Phys. Rev.* A **87** 032312
[26] Simon D S and Sergienko A V 2014 *New J. Phys.* **16** 063052
[27] Fickler R, Lapkiewicz R, Huber M, Lavery M, Padgett M and Zeilinger A 2014 *Interface between Path and OAM Entanglement for High-dimensional Photonic Quantum Information* arXiv:1402.2423 [physics.optics]
[28] Willner A E *et al* 2014 *Adv. Opt. Photon.* **7** 66
[29] Gbur G and Tyson R K 2008 *J. Opt. Soc. Am.* A **25** 225
[30] Gibson G *et al* 2004 *Opt. Express* **12** 5448
[31] Wang J *et al* 2012 *Nat. Photonics* **6** 488
[32] Krenn M, Fickler R, Fink M, Handsteiner J, Malik M, Scheidl T, Ursin R and Zeilinger A 2014 *New J. Phys.* **16** 113028
[33] Ren Y W *et al* 2016 *Opt. Lett.* **41** 622
[34] Krenn M *et al* 2016 *Proc. Natl. Acad. Sci. USA* **113** 13648
[35] Bouchard F *et al* 2018 *Underwater Quantum Key Distribution in Outdoor Conditions with Twisted Photons* arXiv:1801.10299 [quant-ph]
[36] Führhapter S, Jesacher A, Bernet S and Ritsch-Marte M 2005 *Opt. Exp.* **13** 689
[37] Führhapter S, Jesacher A, Maurer C, Bernet S and Ritsch-Marte M 2007 *Adv. Imaging Electron Phys.* **146** 1
[38] Maurer C, Jesacher A, Führhapter S, Bernet S and Ritsch-Marte M 2008 *J. Microsc.* **230** 134
[39] Larkin K G, Bone D J and Oldfield M A 2001 *J. Opt. Soc. Am.* A **18** 1862
[40] Jack B, Leach J, Franke-Arnold S, Ritsche-Marte M, Barnett S M and Padgett M J 2009 *Phys. Rev. Lett.* **103** 083602
[41] Torner L, Torres J P and Carrasco S 2005 *Opt. Exp.* **13** 873
[42] Molina-Terriza G, Rebane L, Torres J P, Torner L and Carrasco S 2007 *J. Eur. Opt. Soc.* **2** 07014
[43] Torres J P, Alexandrescu A and Torner L 2003 *Phys. Rev.* A **68** 050301(R)
[44] Simon D S and Sergienko A V 2012 *Phys. Rev.* A **85** 043825
[45] Uribe-Patarroyo N, Fraine A M, Simon D S, Minaeva O M and Sergienko A V 2013 *Phys. Rev. Lett.* **110** 043601
[46] Fitzpatrick C A, Simon D S and Sergienko A V 2015 *Int. J. Quant. Inf.* **12** 1560013
[47] Paterson C 2005 *Phys. Rev. Lett.* **94** 153901

[48] Bazhenov V Y, Vasnetsov M V and Soskin M S 1990 *JETP Lett.* **52** 429
[49] Annett J F 2004 *Superconductivity, Superfluids, and Condensates* (Oxford: Oxford University Press)
[50] Tsubota M 2009 *Contemp. Phys.* **50** 463
[51] Vilenkin A and Shellard E P S 2000 *Cosmic Strings and Other Topological Defects* (Cambridge: Cambridge University Press)
[52] Aitchison I J R and Hey A J G 2012 *Gauge Theories in Particle Physics: A Practical Introduction* 4th edn (Boca Raton, FL: CRC Press)
[53] Vachaspati T 1998 *Contemp. Phys.* **39** 225
[54] Hegedüs R, Åkesson S, Wehner R and Horváth G 2007 *Proc. Roy. Soc.* A **463** 1081
[55] Morandi G 1992 *The Role of Topology in Classical and Quantum Physics* (Berlin: Springer)
[56] Nakahara M 2003 *Geometry, Topology and Physics* 2nd edn (Boca Raton, FL: CRC Press)
[57] Nash C and Sen S 2013 *Topology and Geometry for Physicists* (Mineola, NY: Dover)
[58] Nash C 1992 *Differential Topology and Quantum Field Theory* (London: Academic)
[59] de Gennes P G and Prost J 1993 *The Physics of Liquid Crystals* (Oxford: Clarendon)
[60] Flossmann F, O'Holleran K, Dennis M R and Padgett M J 2008 *Phys. Rev. Lett.* **100** 203902
[61] Soskin M, Denisenko V and Egorov R 2004 *J. Opt. A: Pure Appl. Opt.* **6** S281
[62] Vasil'ev V and Soskin M 2008 *Opt. Commun.* **281** 5527
[63] Saleh B E A and Teich M C 2019 *Fundamentals of Photonics* 3rd edn (Hoboken, NJ: Wiley)
[64] Zhan Q 2009 *Adv. Opt. Photonics* **1** 1
[65] Beckley A M, Brown T G and Alonso M A 2010 *Opt. Exp.* **18** 10777
[66] Beckley A M, Brown T G and Alonso M A 2012 *Opt. Exp.* **20** 9357
[67] Gu Y, Korotkova O and Gbur G 2009 *Opt. Lett.* **34** 2261
[68] Gu Y and Gbur G 2013 *Opt. Lett.* **38** 1395
[69] Freund I 2010 *Opt. Commun.* **283** 1
[70] Freund I 2010 *Opt. Lett.* **35** 148
[71] Dennis M R 2011 *Opt. Lett.* **36** 3765
[72] Freund I 2011 *Opt. Lett.* **36** 4506
[73] Freund I 2011 *Opt. Commun.* **284** 3816
[74] Freund I 2014 *Opt. Lett.* **39** 727
[75] Bauer T, Banzer P, Karimi E, Orlov S, Rubano A, Marrucci L, Santamato E, Boyd R W and Leuchs G 2015 *Science* **347** 964

IOP Publishing

Topology in Optics (Second Edition)
Tying light in knots
David S Simon

Chapter 7

Knotted and braided vortex lines

7.1 Knotted vortex lines

In the previous chapter, optical phase vortices were discussed. At the center of these vortices, the phase of the optical field is undefined and therefore the amplitude must vanish. In two dimensions, these occur at isolated points. In three-dimensions, vortices generally are lines, rather than points. These vortex lines are not only able to close on themselves to form loops, but they can also tangle around themselves before closing to create complicated knots, and multiple vortex knots can be entwined with each other to form links. In order to form knots and links, the light needs to be polarized and highly coherent, since the vortex lines arise by interference of multiple beams.

Knots first entered physics in the 19th century, when William Tate and Lord Kelvin tried to explain the variety and stability of atoms by modeling them as knotted vortex lines in the ether. Different knot configurations would correspond to different species of atoms, and their stability was guaranteed by the topological stability of the knot, assuming that large amounts of energy would be required for the vortex lines to cross each other. The knot model of atoms fell by the wayside after relativity theory and the Michelson–Morley experiment forced the abandonment of the ether. But Tait's work on classifying knots stimulated the study of knots by mathematicians, and it continues to be a thriving area of topology today.

Since the beginning of the surge in research on vortices and singular fields discussed in the last chapter, a great deal of both theoretical and experimental work has been carried out to study knotted and linked vortex lines, both in optics and in other areas [1–16]. For reviews of developments specifically in optics, see [17–19]. Experimentally, fairly high linking and self-linking numbers have been achieved for collections of knotted optical vortices.

Such vortex curves often arise in speckle fields or other situations where multiple beams interfere, and an examination of those interfering fields show that the resulting dark vortex cores are often highly tangled, forming complicated knots and links, often in a highly fractal structure [20]. These dark vortex knots are most often part of static or locally fluctuating interference patterns, and so they do not propagate. A minimum of three interfering beams are necessary to form such knotted configurations.

Knots are characterized by a number of topological invariants such as the Jones polynomial and the Alexander polynomial which we won't discuss here (see [21] for definitions of these), while links can be characterized by the linking number discussed in chapter 5.

7.2 Creating and characterizing knotted vortices

A method for generating exact analytical solutions to the Helmholtz equation that contain knots and links was developed in [4]. In general, knotted field configurations are built as superpositions of Laguerre–Gauss beams, Bessel beams, or other Helmholtz solutions that contain phase singularities.

Calculating the interfering beams needed to experimentally produce a desired knot or link is a difficult computational problem, and requires calculating the wavevector, intensity, and phase of each beam in the superposition. Typically, in experiments the beams have the correct properties imprinted on them by **spatial light modulators** (SLMs) as in figure 7.1.

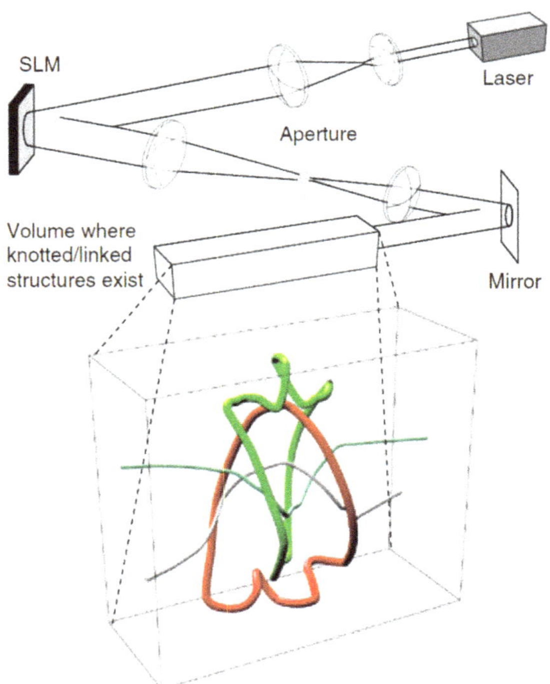

Figure 7.1. Schematic for an experimental arrangement to produce optical vortex knots. (Figures reproduced with permission from [8], copyright 2019 Optical Society of America.)

The SLM has become one of the principle tools in quantum optical research, as well as being found in consumer products like overhead projector systems. An SLM has a set of liquid crystal cells, each of which forms a single pixel of the refleced image. These liquid crystals are nonlinear optical materials whose optical properties can be altered by applying an appropriate voltage; the voltages in turn can be controlled by a computer. The phase and amplitude changes added to light reflecting off each individual pixel can therefore be controllably altered to form complex outgoing optical patterns. In this way, arrays of beams with desired amplitudes, phases, and propagation directions can be produced and interfered with each other in order to form designer light beams with special properties or, more to the point here, vortex lines that are knotted or linked in a desired manner. SLMs can not only create custom tailored output intensity and phase distributions, but they can be controlled in real time allowing dynamically changing configurations to be explored.

The most common types of SLM control only phase, so a common strategy is use the SLM to create a hologram, and then the constructive and destructive interference in the holographic process redistributes the intensity to the desired locations. It has been shown that such a procedure can even work one photon at a time [22], with each photon having amplitudes of travelling in two paths and the knot arising as an interference effect of the photon with itself.

Another means of creating optical vortex knots is via metasurfaces. An **optical metasurface** [23–25] is an artificially engineered surface with subwavelength patterns imprinted on it, for example by an arrangement of attached nanostructures or by variations of the material thickness. Light scattering off the nanoscale patterns allows control of the amplitude, phase, and polarization of the scattered light over extremely fine distance scales, allowing tailoring of the outgoing light with very high spatial resolution. Such metasurfaces are leading to a dramatic miniaturization of many optical devices.

Holographic formation of knots and links on extremely small size scales, up to six orders of magnitude smaller than those created with SLMs, has been carried out using optical metasurfaces [26]. These ultrasmall vortex knots are likely to find application in areas such as the trapping of cold atoms. In addition to phase vortex knots, proposals to use metasurfaces to produce knotted polarization vortices and optical Möbius beams have also appeared [27, 28].

Being able to measure the knotted field and determine its topology is as important as being able to produce it in the first place. The standard means of determining the vortex knot structure experimentally is to image the intensity of multiple cross-sections of the field along the z-axis and then to stack the two-dimensional cross-sections together to form a three-dimensional image. The vortex will then appear as a dark curve threading through the bright light field. Rather than directly imaging each cross-section, another method to reconstruct phase vortices is to interfere the field with a reference plane-wave field on a sequence of planes (figure 7.2). At each of the vortex locations, a fork-shaped discontinuity will appear in the interference pattern.

An improved method has been proposed that increases both the speed and the accuracy of the imaging [29]. Measurements based on digital holography of the

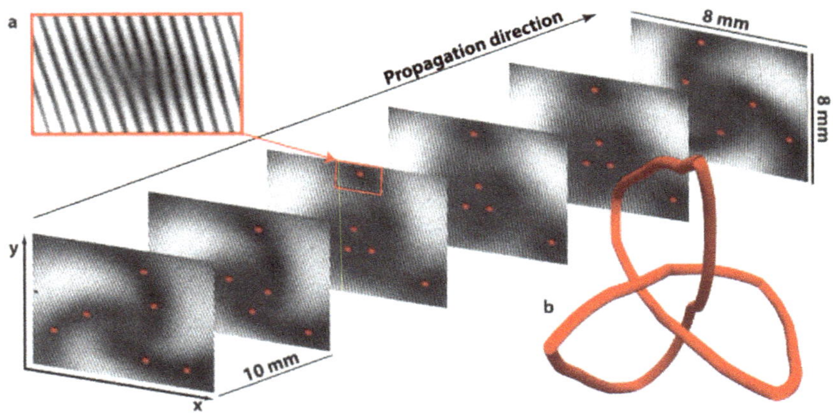

Figure 7.2. Trefoil knot imaged by interfering the field with a plane wave in a sequence of planes. The vortices in each plane are found by looking for the distinctive forked pattern at the location of each vortex, as shown in the inset. (Figures reproduced from [22], copyright 2016 by Authors under Creative Commons License.)

phase distribution in the plane are combined with a numerical search algorithm to efficiently find the singular points. This method holds promise for vector beams as well, and so it should work for polarization vortex knots as well as phase vortex knots.

It can be noted that if the particular knot or link that one wants to detect is known, then its presence or absence can simply be detected by using a hologram. Suppose that a hologram is designed to convert an input Gaussian beam into a particular knot or link. Then light from that knot or link can be converted back into the original Gaussian beam by the same hologram, which can then be detected, for example, by passing it to a detector through a single mode fiber that propagates that particular Gaussian mode [30].

7.3 Variations and applications

In this chapter, the main topic has been knotted vortex lines, where the vortex consists of the set of zero-intensity points of the field. In [31], a holographic method was developed for producing a *bright* optical knot, formed by points of intensity maxima, and for using these bright knots as optical traps for microparticles.

A **freestyle laser trap** [32, 33] uses radiation pressure and phase-gradient forces to trap multiple particles and to propel them along a prescribed path. The path is determined by the trajectory of a bright optical beam. Building on the work with bright knots mentioned above, combined with freestyle trap methods, a method of dynamically routing multiple particles around reconfigurable knotted paths was developed in [34], and was demonstrated by propelling multiple 1 μm silica spheres along knotted paths in a colloidal dielectric; see figure 7.3. This provides a versatile and programmable particle delivery system in three-dimensions with a range of applications in optofluidics, biophysics, and other areas.

One aspect of quantum mechanics that has come to hold a prominent role in recent years, in terms of both fundamental theory and applications, is **entanglement**.

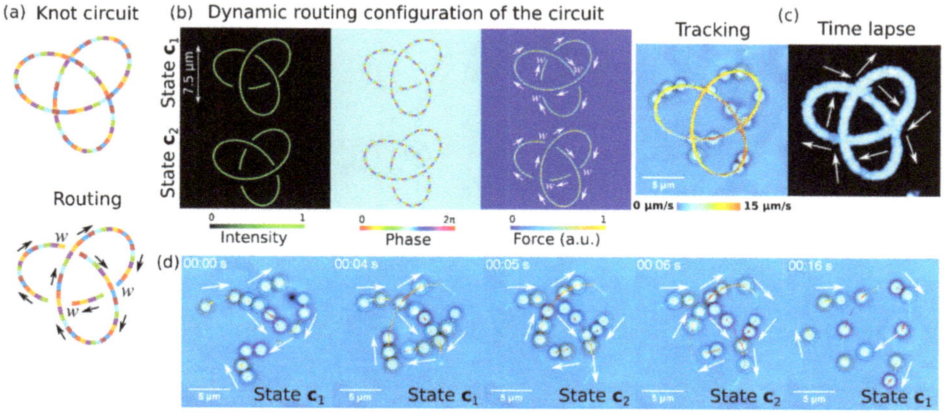

Figure 7.3. Panels (a)–(c) show a bright-field optical trefoil knot. Panels (d)–(f) show experimental results of multiple microparticles being guided and propelled along the knotted path. (Figures reproduced with permission from [34], copyright 2018 Optical Society of America.)

Two particles are said to be entangled if the corresponding two-particle wavefunction cannot be factored into a pair of separate, independent single-particle wavefunctions, $\Psi(x_1, x_2) \neq \psi_1(x_1)\psi_2(x_2)$, where x_1, x_2 are the positions of particles 1 and 2. Entanglement leads to correlations that are in a sense much stronger than any classical correlation, and it is at the heart of many modern applications of quantum mechanics in areas like precision measurement, quantum cryptography, and quantum computing [35–38]. So the entanglement of pairs of knots and links could have significant implications for topological computing (section 11.5).

Recall that knotted and linked structures are often produced by the speckle patterns produced when light reflects off an irregular surface. Also recall that the two photons produced in spontaneous parametric downconversion (chapter 2) are highly entangled with each other and serve as the main source of entangled particles in many quantum optics experiments. The joint two-photon wavefunction of the pair enjoys the high coherence of the original laser source, although each of the two beams individually is of low coherence. Light from laser speckle is often used to produce classical analogs (with reduced correlation properties and lower interference visibility) of quantum effects otherwise produced with downconversion [39]. As a result, it would be reasonable to think that downconverted light could produce knots and links, and that furthermore, the two output beams of the downconversion should be able to produce quantum entangled knots and links.

That this is the case was shown in [30], when parametric downconversion was used to produce a pair of entangled optical Hopf links. A Hopf link is the simplest possible link, just consisting of a pair of interlinked circles. To produce the linked vortices, the downconverted light was imaged onto the surfaces of a pair of SLMs on which computer-generated holograms were displayed. The appearance of the link can be verified by sending the image back to the same holograms to recover the phase and intensity pattern of the light (see figure 7.4). If, instead of returning to the hologram the light in the two beams are sent to coincidence detectors (a pair of

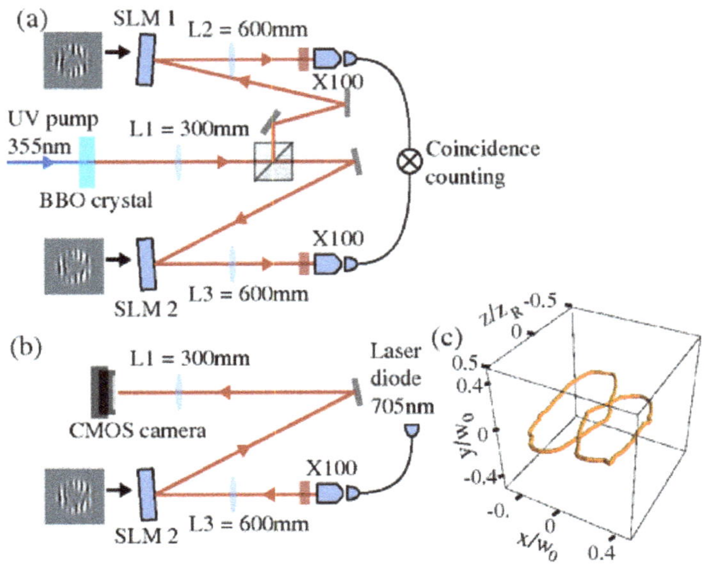

Figure 7.4. (a) The setup for production of a pair of entangled vortices forming a Hopf link and measuring its correlations via coincidence measurement. (b) Sending the image back through one arm of the same setup to verify the link's topology. (c) The recovered image of the link. (Figures reproduced with permission from [30], copyright 2019 Optical Society of America.)

detectors that are linked so a signal is registered only if both detectors fire simultaneously), then the correlations between the two beams can be measured in order to verify that the links are entangled. Specifically, the Clauser–Horne–Shimony–Holt (CHSH) inequality [40] (a variation on the famous quantum mechanical Bell inequality [41, 42]) was verified.

Rather than being specific to optical fields, highly complex knots and links can occur in the wavefunctions of more general quantum systems as well. Computer simulations show that the probability of a given vortex loop being knotted is proportional to its length and that tangles of knotted and linked vortices appear in random superpositions of eigenfunctions of even simple systems like harmonic oscillators [43].

Finally, it should be mentioned that in this chapter the focus was on static knots and links that remain in a fixed location. But propogating optical vortex links can also be produced. These knots and links form as a type of soliton, an isolating shape-invariant propagating wave pulse. Solitons are the subject of the next chapter.

References

[1] Faddeev L and Niemi A J 1997 *Nature* **387** 58
[2] Battye R A and Sutcliffe P M 1998 *Phys. Rev. Lett.* **81** 798
[3] Samuels D C, Barenghi C F and Ricca R L 1998 *J Low Temp Phys.* **110** 509
[4] Berry M V and Dennis M R 2001 *Proc. Roy. Soc.* A **457** 2251
[5] Berry M V and Dennis M R 2001 *J. Phys. A: Math. Gen.* **34** 8877

[6] Dennis M R 2003 *New J. Phys.* **5** 134
[7] Leach J, Dennis M R, Courtial J and Padgett M J 2004 *Nature* **432** 165
[8] Leach J, Dennis M R, Courtial J and Padgett M J 2005 *New J. Phys.* **7** 55
[9] Dennis M R, King R P, Jack B, O'Holleran K and Padgett M J 2010 *Nat. Phys.* **6** 118
[10] Duan Y-S, Zhao L and Zhang X-H 2007 *Commun. Theor. Phys.* **47** 1129
[11] Proment D, Onorato M and Barenghi C F 2012 *Phys. Rev.* E **85** 8
[12] Kedia H, Bialynicki-Birula I and Peralta-Salas D *et al* 2013 *Phys. Rev. Lett.* **111** 5
[13] Kleckner D and Irvine W M 2013 *Nat. Phys.* **9** 253
[14] Hall D S, Ray M W and Tiurev K *et al* 2016 *Nat. Phys.* **12** 478
[15] Arrayas M, Bouwmeester D and Trueba J L 2017 *Phys. Rep.* **667** 1
[16] O'Holleran K, Dennis M R and Padgett M J 2009 *Phys. Rev. Lett.* **102** 4
[17] Padgett M J, O'Holleran K, King R P and Dennis M R 2011 *Contemp. Phys.* **52** 265
[18] Li P, Guo X, Zhong J, Liu S, Zhang Y, Wei B and Zhao J 2020 *Adv. Phys.* **6** 184353
[19] Simon D S 2020 *A Guided Tour of Light Beams: From Lasers to Optical Knots* (Bristol: IOP Publishing)
[20] O'Halloran K, Dennis M R, Flossmann F and Padgett M J 2008 *Phys. Rev. Lett.* **100** 53902
[21] Kauffman L H 1991 *Knots and Physics* (Singapore: World Scientific)
[22] Tempone-Wiltshire S J, Johnstone S P and Helmerson K 2016 *Sci. Rep.* **6** 6
[23] Yu N F, Genevet P, Kats M A, Aieta F, Tetienne J P, Capasso F and Gaburro Z 2011 *Science* **334** 333
[24] Hsiao H H, Chu C H and Tsai D P 2017 *Small Methods* **1** 1600064
[25] Neshev D and Aharonovich I 2018 *Light Sci. Appl.* **7** 58
[26] Wang L, Zhang W and Yin H 2019 *Adv. Opt. Mater.* **7** 1900263
[27] Huo P, Zhang S, Fan Q, Liu Y and Xu T 2019 *Nanoscale* **11** 10646
[28] Wang E, Niu J, Liang Y, Li H, Hua Y, Shi L and Xie C 2020 *Adv. Opt. Mater.* **8** 1901674
[29] Zhong J, Qi S, Liu S, Li P, Wei B, Guo X, Cheng H and Zhao J 2019 *Opt. Lett.* **44** 3849
[30] Romero J, Leach J, Jack B, Dennis M R, Franke-Arnold S, Barnett S M and Padgett M J 2011 *Phys. Rev. Lett.* **106** 100407
[31] Shanblatt E R and Grier D G 2011 *Opt. Express* **19** 5833
[32] Rodrigo J A and Alieva T 2015 *Optica* **2** 812
[33] Rodrigo J A and Alieva T 2016 *Sci. Rep.* **6** 35341
[34] Rodrigo J A, Angulo M and Alieva T 2018 *Opt. Lett.* **43** 4244
[35] Nielsen M A and Chuang I L 2014 *Quantum Computation and Quantum Information* (Cambridge: Cambridge University Press)
[36] Simon D S, Jaeger G and Sergienko A V 2017 *Quantum Metrology, Imaging,and Communication* (Berlin: Springer)
[37] Duarte F J 2019 *Fundamentals of Quantum Entanglement* (Bristol: Institute of Physics Publishing)
[38] Horodecki R, Horodecki P, Horodecki M and Horodecki K 2009 *Rev. Mod. Phys.* **81** 865
[39] Ferri F, Magatti D, Gatti A, Bache M, Brambilla E and Lugiato L A 2005 *Phys. Rev. Lett.* **94** 183602
[40] Clauser J F, Horne M A, Shimony A and Holt R A 1969 *Phys. Rev. Lett.* **23** 880
[41] Bell J S 1964 *Physics* **1** 195
[42] Bell J S 1987 *Speakable and Unspeakable in Quantum Mechanics* (Cambridge: Cambridge University Press)
[43] Taylor J and Dennis M R 2016 *Nat. Commun.* **7** 12346

IOP Publishing

Topology in Optics (Second Edition)

Tying light in knots

David S Simon

Chapter 8

Optical solitons

8.1 Solitary waves

In 1834, Scottish engineer and architect James Scott Russell was conducting experiments at the Edinburgh and Glasgow Union Canal in order to improve the design of canal boats, when he noticed an odd wave pulse arising in the wake a boat that was suddenly stopped. As he followed the pulse along the canal on horseback, he noticed that its shape remained constant, without any noticeable spread in width, and that the amplitude decreased much more slowly than normal water waves. Russell lost sight of the wave after several miles. He soon reproduced a similar wave in a series of experiments in water tanks. This unusual wave pulse, which Russell referred to as a **wave of translation** is now better known as a **solitary wave** or **soliton**.

Solitary waves were viewed largely as a curiosity until the 1960s when waves exhibiting solitonic behavior were found to occur as solutions of the **Korteweg–de Vries (KdV) equation** [1] and to provide an explanation of puzzling results seen by Fermi and his collaborators a decade earlier [2]. Since then, soliton solutions have been found in a number of other nonlinear differential equations, including the nonlinear Schrödinger equation and the Sine–Gordon equation, and they have become ubiquitous in many areas of physics and engineering. They arise for example in particle physics, cosmology, optics, and solid state physics, and may have a role to play in the functioning of the heart [3] and in neuroscience [4]. In particular, in non-Abelian gauge theories used in particle physics, the soliton solutions act as magnetic monopoles [5–7].

Solitons occur only in nonlinear systems, since nonlinear effects are needed to cancel the natural spreading of the waves as they propagate. Because of the nonlinearity, solitons do not obey a superposition principle; but in exchange, they offer a number of other interesting properties. The two most prominent properties of solitons are:

(i) They are unusually stable, retaining their shape over long distances. In ideal cases, they propagate indefinitely without an increase in width or a loss of amplitude. Each soliton pulse remains localized within a finite volume, similar to a discrete particle like a proton or neutron.

doi:10.1088/978-0-7503-3471-6ch8

(ii) When two solitons collide, there is an interaction between them that extends over a finite region. This leads to a complex wave pattern in the interaction region. But eventually the waves separate, leading to a pair of outgoing waves that asymptotically look identical to the incident solitons. In other words, the solitons scatter and move apart, but retain their identity afterwards.

These particle-like properties (stability, localization, and retention of identity after scattering) are the key to the appeal of solitons in areas like particle physics and optical communication. There are many variations on terminology in the literature, but it is common among some authors to refer to waves that satisfy property (i) as solitary waves and to reserve the name soliton for waves that satisfy both (i) and (ii). Here, we will not make this distinction and will treat the two terms as interchangeable.

Solitons come in two kinds: **topological solitons** can change the values of topological quantum numbers such as the winding number, while **nontopological solitons** preserve all topological indices. In order to conserve topological charge, the soliton itself must carry topological charge equal to the difference between the charges of the initial and final states.

8.2 Simple example: Sine–Gordon equation

To provide a concrete example, consider the so-called **Sine–Gordon equation** [5]. Let $\phi(x, t)$ be a real scalar field in one space dimension (plus time), and suppose that it has a potential energy function given by

$$V(\phi) = 1 - \cos\phi, \tag{8.1}$$

where all dimensional parameters (mass and coupling constant) have been set equal to 1 for simplicity. Using the Euler–Lagrange equation, it is straightforward to find the equation of motion for this field:

$$\frac{\partial^2\phi}{\partial t^2} + \frac{\partial^2\phi}{\partial x^2} + \sin\phi = 0. \tag{8.2}$$

The corresponding energy of the field is found by integrating the energy density over all of the one-dimensional space,

$$E = \int_{-\infty}^{\infty}\left[\frac{1}{2}\left(\frac{\partial\phi}{\partial t}\right)^2 + \frac{1}{2}\left(\frac{\partial\phi}{\partial x}\right)^2 + (1 - \cos\phi)\right] dx. \tag{8.3}$$

The energy has an infinite set of degenerate minima: $E = 0$ whenever the field is temporally static and spatially constant at one of the values $\phi_n = 2\pi n$, $n = 0, \pm1, \pm2, \ldots$. Here we seek solutions that are static and finite energy, but not necessarily spatially constant. The simplest such solution is not hard to find by the method of quadrature [5]: if we take $n = n_0$ at $x = -\infty$, then it can be verified by direct substitution that

$$\phi_{\text{sol}} = \phi_{n_0} + 4\tan^{-1}(e^{x-x_0}) \tag{8.4}$$

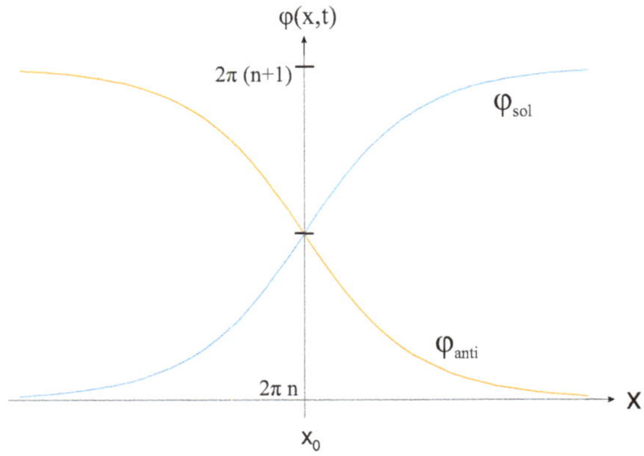

Figure 8.1. Soliton and antisoliton solutions to the Sine–Gordon equation. Each approaches two adjacent zero-energy minima asymptotically as $x \to \pm\infty$.

satisfies the equation of motion for any constant x_0, as does $\phi_{\text{anti}}(x) = \phi_{n_0+1} - 4\tan^{-1}(e^{x-x_0})$. These are the **soliton** and **antisoliton** solutions.

The soliton solution approaches minima as $x \to \pm\infty$: if $\phi(-\infty, t) = \phi_{n_0}$, then $\phi(+\infty, t) = \phi_{n_0+1}$. In other words, the soliton interpolates between two adjacent vacua separated by $\Delta n \equiv n(+\infty) - n(-\infty) = 1$. Similarly, the antisoliton interpolates between minima with decreasing n: $\Delta n = -1$; see figure 8.1.

Defining a winding number or topological charge,

$$Q = \frac{1}{2\pi}\int\left(\frac{d\phi}{dx}\right)dx = \Delta n, \tag{8.5}$$

we clearly have $Q_{\text{sol}} = +1$ and $Q_{\text{anti}} = -1$. We may think of these topological solitons as particle-like excitations of the field which carry a unit of topological charge to the right (soliton) or to the left (antisoliton). Multi-soliton and soliton–antisoliton scattering solutions also exist [5].

Since the solutions are temporally static, we can look at the field at any fixed time. Compactifying the spatial line by identifying $x = +\infty$ with $x = -\infty$, space is effectively a circle, S^1. Similarly, the field, which goes from one multiple of 2π to another as the spatial circle is traversed, can also be thought of as describing motion around a circle $e^{i\phi(x,\,t)}$ in the complex plane. So the integer-valued topological charge is labelling the homotopy class of the map ϕ: $S^1 \to S^1$ defined by the field.

In chapter 10, similar topological solitons will occur as edge or boundary states confined between regions of a solid residing in different topological phases and the soliton will provide an interpolation between the different topological quantum numbers of the phases.

8.3 Solitons in optics

The topic of optical solitons is a large and rapidly expanding one, with applications throughout nonlinear optics and in optic communications systems. A number of reviews exist, including [8–10]. In optics, there is usually a distinction between **spatial** and **temporal solitons**. In the former case, the wave pulse is localized in the transverse spatial dimensions. The usual spatial spread of the optical pulse due to diffraction is compensated by self-focusing in a nonlinear optical material. In the temporal case, the wave may be widely distributed in the transverse direction, but it is strongly localized in the longitudinal direction, with the pulse shape staying constant over time. In this case, the normal temporal pulse broadening due to dispersion is compensated by self-phase modulation. There can also be spatiotemporal solitons, which are commonly referred to as **light bullets**.

The first optical spatial soliton was seen in studies of self-trapping of continuous wave laser beams in nonlinear media [11]. Temporal solitons were observed soon after in experiments on self-induced transparency [12], and were later shown to be capable of propagating in optical fibers [13, 14], opening up the possibility of using them for communication and information processing.

The solitons just mentioned are all bright solitons: they consist of localized high intensity pulses propagating through a darker background. They exist in the presence of self-focusing (chapter 1). However, dark solitons, which occur in the presence of self-defocusing, have also been studied since being observed in the early 1970s [15]. These are localized pulses of darkness in a bright light field, or in other words holes or intensity zeros around which the field may wind. These dark solitons may be pointlike, or may form filaments or rings. As with the vortex lines of chapters 6 and 7, these can become knotted, tangled, and linked. However, there are two significant differences between vortex knots and soliton knots: (i) vortices can occur in free space, whereas solitons require a nonlinear optical medium. (ii) The vortex knots are static interference patterns; they do not propagate. This is in contrast to solitonic knots, which move through space at fixed velocity.

In nonlinear optical media, the effective index of refraction depends on the intensity of the light. Consider a Kerr medium, where the polarization nonlinearity is cubic, $\mathcal{O}(E^3)$, as in section 2.3, and send a light beam through it. Assume for the moment that the material is two dimensional: one longitudinal direction z and one transverse x. Begin with the nonlinear Schrödinger equation, equation (2.49), and attempt a trial solution of the form $E(x, z, t) = A(x, z, t)e^{ikz}$. Going over to dimensionless coordinates, $x' = x/w_0$, $y' = y/w_0$, and $z' = z/z_0$, where w_0 is the beam width and z_0 is the Rayleigh range, we may also define a dimensionless field amplitude u

$$u\sqrt{I_0} = A, \tag{8.6}$$

where I_0 is the maximum intensity. The nonlinear Schrödinger equation for a self-focusing medium can then be put into the form

$$i\frac{\partial u}{\partial z'} + \frac{1}{2}\frac{\partial^2 u}{\partial x'^2} - |u|^2 u = 0. \tag{8.7}$$

It can be verified by direct substitution that one solution to this equation is

$$u(x', z') = a\ \mathrm{sech}(ax')e^{ia^2z'/2}, \tag{8.8}$$

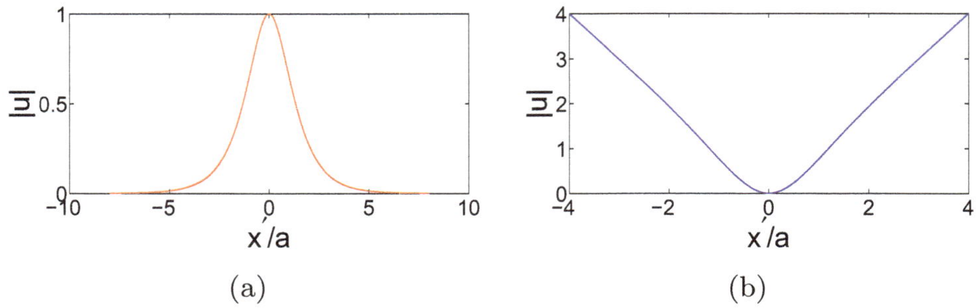

Figure 8.2. Intensity profiles in the transverse direction for examples of (a) a bright soliton ($I \sim |\text{sech}(ax)|^2$) and (b) a dark soliton ($I \sim |x \tanh(ax)|^2$), for the case of one transverse direction.

where a is a constant playing the role of the soliton amplitude. This corresponds to a dimensional field amplitude of

$$A(x, z) = I_0; \text{sech}\left(\frac{x}{w_0}\right)e^{iz/2z_0}. \tag{8.9}$$

This solution, which is plotted in figure 8.2(a), represents a bright soliton. Note that $|A(x, z)|$ is actually independent of z, so the pulse, which is localized within a distance $\sim 1/a$ in the x direction, retains its spatial shape as it propagates.

In a similar manner, a dark soliton of the form

$$u(x', z') = x' \tanh(ax')e^{-ia^2z'}, \tag{8.10}$$

exists in a self-defocusing medium ($n_2 < 0$) (figure 8.2(b)). Note the intensity zero at the center, characteristic of a phase singularity.

Going from one to two transverse directions, the two-dimensional nonlinear Schrödinger equation in dimensionless form is

$$i\frac{\partial u}{\partial z'} + \frac{1}{2}\left(\frac{\partial^2 u}{\partial x'^2} + \frac{\partial^2 u}{\partial y'^2}\right) - |u|^2 u = 0. \tag{8.11}$$

In a self-defocusing material, a change of variable called a **Madelung transform** [16, 17] can be used to write the complex electric field in terms of a new dimensionless field in cylindrical coordinates, $\psi(r, \theta, z)$ (see [8] for details), that solves the nonlinear Schrödinger equation. Attempting a solution of the form

$$\psi(r, \theta, z) = \chi(r)e^{im\theta}, \tag{8.12}$$

such a solution works as long as χ satisfies

$$\frac{d^2\chi}{dr^2} + \frac{1}{r}\frac{d\chi}{dr} - \frac{m^2}{r^2}\chi + (1 - \chi^2)\chi = 0. \tag{8.13}$$

This gives a singular phase vortex solution with topological charge m and OAM $m\hbar$. The phase singularity forces $\chi(0) = 0$, so taking a boundary condition that $\chi(r)$ approaches a constant as $r \to \infty$, χ has asymptotic forms

$$\chi(r) \sim \begin{cases} ar^{|m|} & \text{for } r \to 0, \\ 1 - \dfrac{m^2}{2r^2} & \text{for } r \to \infty. \end{cases} \tag{8.14}$$

As with the Laguerre–Gauss OAM vortices discussed in chapter 6, these dark soliton vortices tend to be unstable for $|m| > 1$, breaking up into multiple vortices of $|m| = 1$ under environmental perturbations [18, 19].

There are other dark solitons that can be created in self-defocusing media. Dark soliton lines or stripes can be formed, but they tend to be strongly unstable unless the ends are joined together to form closed rings [20]. Such rings are structurally stable, but tend to grow in radius as they propagate, and so they will miss any finite-size detector if allowed to propagate sufficient distance. The only stationary, fixed size solutions that remain are the solitonic vortex solutions described above. The idea that solitonic structures could form into loops and become knotted was explored as far back as the 1970s [21, 22]

If a plane wave hits one of these vortex solitons, the wavefront will split into two outgoing wavefronts, one on each side of the vortex and shifted relative to each other in the longitudinal direction by a distance of $2\pi/v$, where v is the wave speed. This shift in longitudinal position across the vortex is a result of a phase difference between the two sides. This is due to a geometric phase and is directly related to the Aharonov–Bohm effect [23] (figure 8.3).

The solitons described so far in this chapter live in conservative systems, in which the energy is constant. However, solitons may also appear in dissipative systems as well [24–27], and can form quite complicated structures, with nontrivial topology such as knotted and linked structures (figure 8.4). Unlike the knots and link of the previous chapter, these solitonic solutions will propagate through space.

We have only discussed optical solitons based on Kerr nonlinearities, but other classes of solitons exist. For example, **photorefractive solitons** [28–31] rely on the nonuniform illumination of a material, causing charge generation due to light absorption and leading to a spatially dependent alteration of the refractive index. This has the advantage that the optical power required is much lower than required for Kerr-based solitons. Another possibility is the **quadratic soliton** [32–36], which is

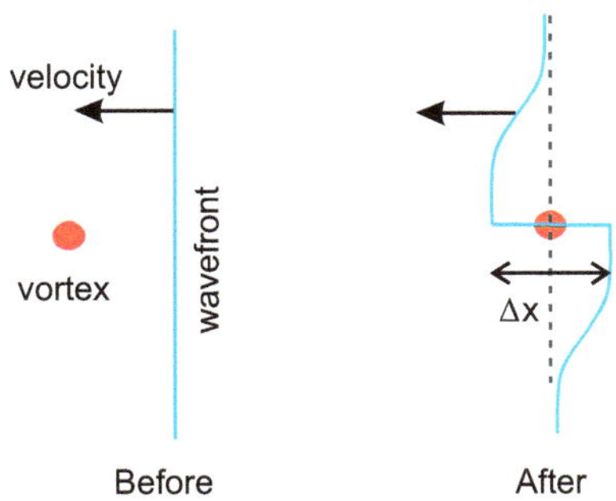

Figure 8.3. The appearance of an Aharonov–Bohm phase shift leads to a wavefront becoming split into two longitudinally displaced pieces after encountering a vortex soliton.

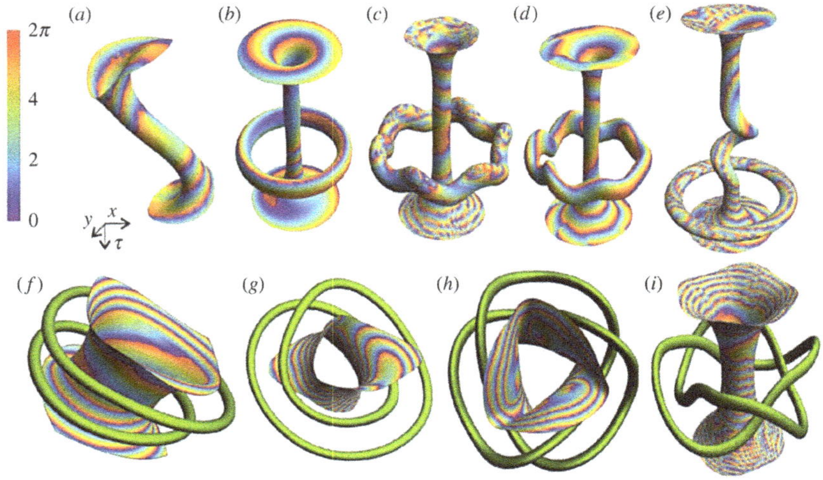

Figure 8.4. Vortex solitons in dissipative systems, including examples that are knotted (h) and linked (i). (Figures reproduced with permission from [26], copyright 2018 the authors.)

based on second harmonic generation (i.e. on the χ_2 term in equation (2.40), rather than on the χ_3 Kerr term).

Clearly, we have only exposed the tip of the very large iceberg presented by optical solitons. The ability of these wave pulses to propagate without distortion through nonlinear media opens up a range of applications that are only starting to be explored. For example, it has been proposed that dark ring solitons can form waveguides to maintain the stability of clusters of multiple bright solitons [37, 38], which could be useful in communications applications. In chapter 10, a different means of transmitting optical states without distortion will be presented; instead of using the *local* properties of nonlinear materials, the approach presented there will require tailoring the *global* structure of the propagation medium to produce topologically based effects.

References

[1] Zabusky N J and Kruzkal M D 1965 *Phys. Rev. Lett.* **15** 240

[2] Fermi E, Pasta J and Ulam S 1955 *Studies of Nonlinear Problems* (Los Alamos, NM: Los Alamos National Laboratory) Document LA-1940

[3] Aslanidi O V and Mornev O A 1999 *J. Biol. Phys.* **25** 149

[4] Andersen S, Jackson A and Heimburg T 2009 *Prog. in Neurobio.* **88** 104

[5] Rajaraman R 1987 *Solitons and Instantons: An Introduction to Solitons and Instantons in Quantum Field Theory* (Amsterdam: North-Holland)

[6] Ryder L H 1996 *Quantum Field Theory* 2nd edn (Cambridge: Cambridge University Press)

[7] Cheng T P and Li L F 1988 *Gauge Theory of Elementary Particle Physics* (Oxford: Oxford University Press)

[8] Kivshar Y S and Agrawal G P 2003 *Optical Solitons: From Fibers to Photonic Crystals* (San Diego, CA: Academic)

[9] Taylor J R 2005 *Optical Solitons: Theory and Experiment* (Cambridge: Cambridge University Press)

[10] Chen Z, Segev M and Christodoulides D N 2012 *Rep. Prog. Phys.* **75** 086401
[11] Chaio R Y, Garmire E and Townes C H 1964 *Phys. Rev. Lett.* **13** 479
[12] McCall S L and Hahn E L 1967 *Phys. Rev. Lett.* **18** 908
[13] Hasegawa A and Tappert F 1973 *Appl. Phys. Lett.* **23** 142
[14] Hasegawa A and Tappert F 1973 *Appl. Phys. Lett.* **23** 171
[15] Kivshar Y S and Luther-Davies B 1998 *Phys. Rep.* **298** 81
[16] Madelung E 1927 *Zeit. F. Phys.* **40** 322
[17] Delphenich D H 2002 *The Geometric Origin of the Madelung Potential* arXiv:gr-qc/0211065
[18] Dreischuh A, Paulus G G, Zacher F, Grabson E, Neshev D and Walther H 1999 *Phys. Rev.* E **60** 7518
[19] Firth W and Skryabin D 1997 *Phys. Rev. Lett.* **79** 2450
[20] Kuznetsov E A and Turitsyn S K 1988 *Sov. Phys. JETP* **67** 1583
[21] Faddeev L D 2005 *Quantization of solitons* Princeton preprint IAS-75-QS70 (Princeton, NJ: Institute for Advanced Study)
[22] Faddeev L 2001 *Phil Trans. Math. Phys. Eng. Sci.* **359** 1399
[23] Neshev D, Nepomnyashchy A and Kivshar Yu S 2001 *Phys. Rev. Lett.* **87** 043901
[24] Veretenov N A, Rosanov N N and Fedorov S V 2016 *Phys. Rev. Lett.* **117** 183901
[25] Veretenov N A, Fedorov S V and Rosanov N N 2017 *Phys. Rev. Lett.* **119** 263901
[26] Veretenov N A, Fedorov S V and Rosanov N N 2017 *Phil. Trans. R. Soc.* A **376** 20170367
[27] Rosanov N N 2011 *Dissipative optical solitons* (Moscow: Fizmatlit)
[28] Segev M, Crosignani B, Yariv A and Fischer B 1992 *Phys. Rev. Lett.* **68** 923
[29] Duree G C *et al* 1993 *Phys. Rev. Lett.* **71** 533
[30] Castillo M D I, Aguilar P A M, Sanchez-Mondragon J J, Stepanov S and Vysloukh V 1994 *Appl. Phys. Lett.* **64** 408
[31] Crosignani B, Di Porto P, Segev M, Salamo G and Yariv A 1998 *Riv. Nuovo Cimento* **21** 1
[32] Hayata K and Koshiba M 1993 Multidimensional solitons in quadratic nonlinear media *Phys. Rev. Lett.* **71** 3275–8
[33] Torruellas W E, Wang Z, Hagan D J, VanStryland E W, Stegeman G I, Torner L and Menyuk C R 1995 *Phys. Rev. Lett.* **74** 5036
[34] Schiek R, Baek Y and Stegeman G I 1996 *Phys. Rev.* E **53** 1138
[35] Stegeman G I, Hagan D J and Torner L 1996 *Opt. Quantum Electron.* **28** 1691
[36] Buryak A V, Di Trapani P, Skryabin D V and Trillo S 2002 *Phys. Rep.* **370** 63
[37] Dreischuh A, Kamenov V and Dinev S 1996 *Appl. Phys.* B **63** 145
[38] Sheppard A E and Kivshar Y S 1997 *Phys. Rev.* E **55** 4773

Chapter 9

Geometric and topological phases

When the parameters describing a classical system return to their original value, the system normally returns to its initial state. This is not necessarily true in quantum mechanics, where the wavefunction can gain a path-dependent phase; such phases are detectible via interference and often lead to physically significant effects. These phases are distinct from the dynamical phase factors e^{-iEt} introduced by time evolution and are ultimately of geometric or topological origin.

Such phases became well-known after the work of Michael Berry in the 1980s [1], studying cyclic, adiabatic parameter changes in quantum systems, but in fact they had made occasional appearances in physics much earlier. In particular, Pancharatnam [2] discovered a related effect while looking at polarization in a classical optical system.

The phases of interest here arise from the holonomy on the closed path C taken by a set of parameters or control variables (polarization, magnetic field orientation, etc) in some parameter space; this space may be of any dimension, N, unrelated to the dimension of the three-dimensional physical space. The phases are **geometric phases** if they depend on the shape of the *individual* path (and in particular on the angle subtended by the path with respect to some point); in contrast, they are **topological phases** if they depend only on the topological equivalence class of the curve, but are independent of smooth path variations within that class.

The importance of these nondynamical phases is reflected in the fact that many variations on them have appeared in many areas of physics, under a variety of different names (including Berry phases, Pancaharatnam phases, geometric phases, topological phases, Zak phases, and Hannay angles, among others). The abundance of variations is made more confusing by the fact that different authors often use mutually inconsistent terminology for these phases. To avoid this, we will generically refer to such nondynamical phases as geometric phases or Berry phases, regardless of their origin. In particular, we avoid the use of the term *topological phase*, since that term is also used for a completely different concept that will be discussed in chapter 10.

doi:10.1088/978-0-7503-3471-6ch9

A brief guide to the major developments in the history of geometric phases can be summarized as follows. After the initial papers of Pancaharatnam [2] in polarization optics and Berry [1] on quantum mechanics of nondegenerate systems, Wilczek and Zee [3] generalized to the case of degenerate systems with higher dimensional ground states. This gave rise the non-Abelian geometric phase. Aharonov and Anandan [4] removed Berry's assumption of adiabaticity (slow variation of the parameters), while Samuel and Bhandari [5] looked at non-cyclic variations. In quantum mechanics, Uhlmann [6] generalized from pure quantum states to mixed quantum states. Zak [7] looked at electrons in solids and took the cyclic parameter variation to be the variation of quasi-momentum across a Brillouin zone; these Zak phases and their optical analogs will come up in chapters 10 and 11. Meanwhile, Hannay [8] studied analogous phases in classical mechanical systems, showing, for example, that they play a role in the behavior of the Foucault pendulum. On the mathematical front, Simon [9] showed that geometric phases are physical manifestations of holonomy on fiber bundles, as discussed in chapter 4. A more detailed guide to the literature related to these topics may be found in [10].

9.1 The Pancharatnam phase

Consider interference between two light beams with electric fields

$$E_a = |E_a|e^{i\phi_a} \tag{9.1}$$

$$E_b = |E_b|e^{i(\phi_a+\delta\phi)}, \tag{9.2}$$

where $\delta\phi = \phi_b - \phi_a$. Assuming no dissipation, the magnitudes of the two fields don't change as they interfere, so without loss of generality we can rescale both fields by a common constant factor in order to normalize them:

$$|E_a|^2 + |E_b|^2 = 1, \tag{9.3}$$

which means in turn that the field amplitudes can be parameterized by an angular variable,

$$|E_a| = \cos\frac{\theta}{2}, \qquad |E_b| = \sin\frac{\theta}{2}. \tag{9.4}$$

In 1956, Pancharatnam, in the course of investigating the polarization of light, asked the following question: How can the phase difference between two different but non-orthogonal polarization states of light be defined? Consider the two beams of light above with non-orthogonal polarization states, represented by points a and b on the Poincaré sphere (figure 2.6). Let $\theta_{ab}/2$ be the angle between the polarizations in space; then the angle between the corresponding states on the Poincaré sphere is θ_{ab}. The fields and intensities of these beams are $I_a = |\boldsymbol{E}_a|^2$, $I_b = |\boldsymbol{E}_b|^2$. Pancharatnam's first step was to break $\boldsymbol{E}_b$ into components parallel and perpendicular to $\boldsymbol{E}_a$: $\boldsymbol{E}_b = \boldsymbol{E}_\parallel + \boldsymbol{E}_\perp$, with intensities $I_\parallel = |E_b|^2\cos^2\theta_{ab}/2$ and $I_\perp = |E_b|^2\sin^2\theta_{ab}/2$. There is no interference between orthogonal polarization components, so the phase difference $\delta\phi$ between the beams is defined to be the phase difference between the parallel components, $\boldsymbol{E}_a$ and $\boldsymbol{E}_{\|}$, which interfere to give an intensity

$$\begin{aligned} I' &= (\boldsymbol{E}_a + \boldsymbol{E}_{\|})^* \cdot (\boldsymbol{E}_a + \boldsymbol{E}_{\|}) \\ &= I_a + I_b \cos^2 \frac{\theta_{ab}}{2} + 2\sqrt{I_a I_b} \cos \frac{\theta_{ab}}{2} \cos \delta\phi. \end{aligned} \tag{9.5}$$

Adding in the component of b perpendicular to a, the total intensity is then

$$I = I' + I_b \sin^2 \frac{\theta_{ab}}{2} = I_a + I_b + 2\sqrt{I_a I_b} \cos \frac{\theta_{ab}}{2} \cos \delta\phi. \tag{9.6}$$

The phase difference between the two differently polarized optical states is then defined to be the phase by which one beam must be retarded or advanced to make the interference between the two beams reach maximum intensity.

Suppose that the polarization state of the combined beam is represented by point c on the Poincaré sphere (figure 9.1) Now take the points a' and b', directly opposite to a and b on the Poincaré sphere, representing the optical polarization states orthogonal to a and b. Define $\theta_{a'c}$ and $\theta_{b'c}$ to be the angles between a' and c and between b' and c on the sphere. Then, it can be shown [2, 11] that the relative phase angle $\delta\phi$ is related to the angles on the sphere by

$$-\cos \delta\phi = \frac{-1 + \cos^2 \frac{\theta_{ab}}{2} + \cos^2 \frac{\theta_{a'c}}{2} + \cos^2 \frac{\theta_{b'c}}{2}}{2 \cos^2 \frac{\theta_{ab}}{2} \cos^2 \frac{\theta_{b'c'}}{2} \cos^2 \frac{\theta_{a'c}}{2}}. \tag{9.7}$$

Pancharatnam then noticed that the quantity on the right side of this formula can be written much more simply in terms of the solid angle Ω subtended by the geodesic triangle $a'b'c$ (figure 9.1(c)) about the center of the Poincaré sphere:

$$\cos \delta\phi = -\cos \frac{\Omega}{2}. \tag{9.8}$$

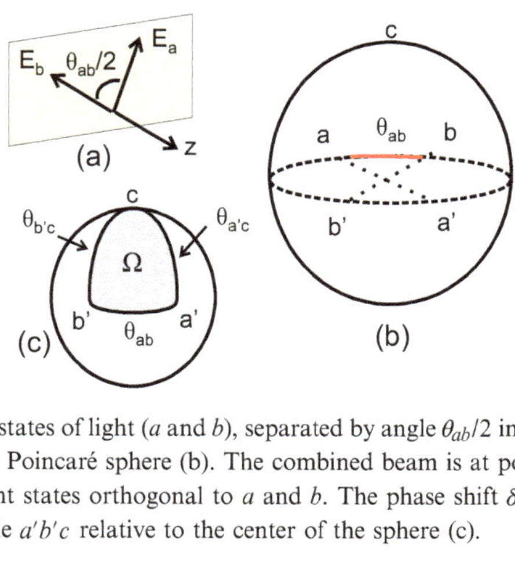

Figure 9.1. Two polarization states of light (a and b), separated by angle $\theta_{ab}/2$ in space (a) are located at points separated by angle θ_{ab} on the Poincaré sphere (b). The combined beam is at point c on the sphere, while the points a' and b' represent light states orthogonal to a and b. The phase shift $\delta\phi$ is related to the solid angle subtended by geodesic triangle $a'b'c$ relative to the center of the sphere (c).

In other words, the phase difference between the beams is given by

$$|\delta\phi| = \pi - \frac{\Omega}{2}. \tag{9.9}$$

This phase is clearly of geometric origin, being given by the size of a solid angle. It is the first known explicit appearance of a geometric phase in physics. The Pancharatnam phase is a geometric phase acquired along paths through the Hopf bundle (see section 4.7 or [12, 13]). When discussing polarization, the magnitude of the electric field is irrelevant; only the direction and the relative phases of the components matter. So, the possible polarization states are spanned by a complex spinor of the form

$$z = \begin{pmatrix} z_1 \\ z_2 \end{pmatrix} = \begin{pmatrix} \cos\frac{\theta}{2}\, e^{i\phi_a} \\ \sin\frac{\theta}{2}\, e^{i(\phi_a+\delta\phi)} \end{pmatrix}, \tag{9.10}$$

where the components $z_j = x_j + iy_j$ are normalized to

$$|z_1|^2 + |z_2|^2 = x_1^2 + y_1^2 + x_2^2 + y_2^2 = 1. \tag{9.11}$$

These components span a three-dimensional sphere, S^3. The Hopf bundle has this S^3 as the total space, and the Poincaré sphere as the base. The fiber is the $U(1)$ group, which again is topologically equivalent to a circle, S^1. Physically, the fiber variable is the overall phase of the spinor; this is irrelevant to the polarization state described by the Poincaré base manifold at a given moment, but is needed in order to compare the polarizations at different points via interference. The change in this overall phase along some path is the Pancharatnam phase.

The normalized Stokes vector $\boldsymbol{s} = \boldsymbol{S}/|\boldsymbol{S}|$ is given by

$$s_1 = \sin\theta\cos\delta\phi \tag{9.12}$$

$$s_2 = \sin\theta\sin\delta\phi \tag{9.13}$$

$$s_3 = \cos\theta. \tag{9.14}$$

This Stokes vector is obtained from the polarization spinor via the Hopf projection,

$$\pi\colon z \to \boldsymbol{s}, \qquad \boldsymbol{s} = z^\dagger \boldsymbol{\sigma} z. \tag{9.15}$$

The parallel transport condition along a curve C in the bundle can be written in the form

$$z^\dagger\, dz = 0, \tag{9.16}$$

or equivalently as $A(\dot{z}) = 0$, where $A = -iz^\dagger\, dz$ is the connection one-form, and $\dot{z}$ is the tangent vector along the curve C. (Note that, for convenience, we have reordered the components $s_1 \to s_2 \to s_3 \to s_1$ relative to the definitions in section 2.4.)

In the next section, we examine geometrical phases in a broader context, seeing how they arise in a generic quantum system. In place of an overall phase of the Jones vector, the Berry phase will be an overall phase attached to a quantum state vector.

9.2 Berry phase in quantum mechanics

The state of a quantum mechanical system is specified by a state vector $|\psi\rangle$ in a Hilbert space. More precisely, the state is defined by a ray of state vectors in the Hilbert space, since $|\psi\rangle$ and $C|\psi\rangle$ represent the same state for any complex constant, C. As a result, the magnitude of C is usually fixed by normalizing to $|C|^2 = 1$. Note that this still leaves the phase of C undetermined. As long as C is constant for all vectors in the Hilbert space, it too is irrelevant and can be fixed arbitrarily. However, the difference between phases of different states is important, so any part of the phase that is not constant must be retained. We focus on this phase in the remainder of this section.

The inner product of two normalized states, $\langle\psi_1|\psi_2\rangle$, is in general a complex number, and so can be written in polar form as

$$\langle\psi_2|\psi_1\rangle = Ae^{i\phi}, \tag{9.17}$$

where $A = |\langle\psi_2|\psi_1\rangle|$ and ϕ are real numbers. ϕ is interpreted as the phase difference between the two states. So we can *define* the phase difference between a pair of states by solving the latter equation for ϕ:

$$\phi = -i \text{ Im} \ln\left(\frac{\langle\psi_2|\psi_1\rangle}{|\langle\psi_2|\psi_1\rangle|}\right). \tag{9.18}$$

These relative phases are not gauge invariant; under a change of phase of each state, $|\psi_j\rangle \to e^{-i\delta\phi_j}|\psi_j\rangle$, the relative phase changes according to

$$\phi \to \phi + (\delta\phi_2 - \delta\phi_1). \tag{9.19}$$

Now imagine a sequence of discrete transitions between N states. After reaching $|\psi_N\rangle$, the system returns to the initial state $|\psi_1\rangle$, forming a closed cycle (figure 9.2). Despite returning back to the initial state, the phase gained around this closed cycle is in general nonzero:

$$\phi = -i \text{ Im} \ln\left(\frac{\langle\psi_1|\psi_N\rangle \cdots \langle\psi_3|\psi_2\rangle\langle\psi_2|\psi_1\rangle}{|\langle\psi_1|\psi_N\rangle \cdots \langle\psi_3|\psi_2\rangle\langle\psi_2|\psi_1\rangle|}\right). \tag{9.20}$$

Note that the phase change around this full cycle *is* gauge invariant, since each of the $\delta\phi_i$ cancels between two adjacent terms in the product, implying that ϕ is a physically measurable quantity.

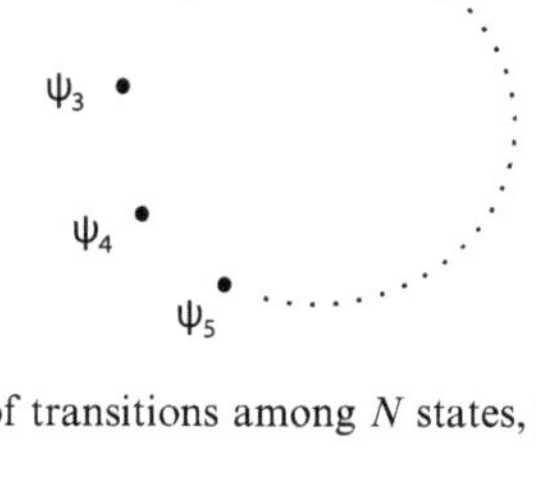

Figure 9.2. A cyclic sequence of transitions among N states, $|\psi_1\rangle \to |\psi_2\rangle \to \cdots \to| \psi_N\rangle \to |\psi_1\rangle$.

Rather than a discrete sequence of states, consider now a continuous path in the Hilbert space. Once again such a phase will appear; let us examine it in more detail. Suppose that the system depends on some set of continuous parameters (temperature, magnetic field, etc). If there are d of these parameters, we may form a d-dimensional vector $\boldsymbol{\lambda}$ from them. We also define the gradient in the abstract parameter space, $\nabla_\lambda = \{\partial_{\lambda_1}, \partial_{\lambda_2}, \ldots \partial_{\lambda_d}\}$. As the parameters are smoothly varied over time, the state of the system will trace out a continuous path in the Hilbert space. For example, polarization states or spin states will trace out paths on the corresponding Poincaré or Bloch sphere. The Hamiltonian will also be a function of the parameters, and the Schrödinger equation will take the form

$$H(\boldsymbol{\lambda}(t))|\psi(\boldsymbol{\lambda}(t))\rangle = i\hbar\frac{d}{dt}|\psi(\boldsymbol{\lambda}(t))\rangle \tag{9.21}$$

The eigenstates will be labeled by a discrete label n,

$$H(\boldsymbol{\lambda})|\psi_n(\boldsymbol{\lambda})\rangle = E_n|\psi_n(\boldsymbol{\lambda}), \rangle \tag{9.22}$$

with $\langle\psi_m|\psi_n\rangle = \delta_{mn}$. We assume for simplicity that the eigenstates are nondegenerate. If the variation of parameters is sufficiently slow, the adiabatic theorem of quantum mechanics [14–16] implies that a state in the nth level will remain in the nth level even as the levels gradually move with changing system parameters; there will be no sudden jumps to other levels.

Let the system initially be in the nth eigenstate: $|\psi(0)\rangle = |\psi_n(\boldsymbol{\lambda}(0))\rangle$. Then it should remain in the nth (slowly changing) state as time passes; however, there can be an unknown phase change as the parameters evolve:

$$|\psi(t)\rangle = e^{-i\theta(t)}|\psi_n(\boldsymbol{\lambda}(t))\rangle. \tag{9.23}$$

To determine the unknown phase, plug the state of equation (9.23) into the Schrödinger equation. After canceling the exponential term from both sides, we find

$$E_n(\boldsymbol{\lambda}(t))|\psi_n(\boldsymbol{\lambda}(t))\rangle = \hbar\left[\frac{d\theta(t)}{dt} + i\frac{d}{dt}\right]|\psi_n(\boldsymbol{\lambda}(t))\rangle. \tag{9.24}$$

Taking the inner product of both sides with $\langle\psi_n(t)|$,

$$E_n(\boldsymbol{\lambda}(t)) = \hbar\frac{d\theta(t)}{dt} + i\hbar\langle\psi_n(\boldsymbol{\lambda}(t))|\frac{d}{dt}|\psi_n(\boldsymbol{\lambda}(t))\rangle, \tag{9.25}$$

or:

$$\hbar\frac{d\theta(t)}{dt} = E_n(\boldsymbol{\lambda}(t)) - i\hbar\langle\psi_n(\boldsymbol{\lambda}(t))|\frac{d}{dt}|\psi_n(\boldsymbol{\lambda}(t))\rangle. \tag{9.26}$$

Integrating both sides, the result is

$$\theta(t) = \frac{1}{\hbar}\int_0^t E_n(\lambda(t'))dt' - i\int_0^t \langle\psi_n(\lambda(t'))|\frac{d}{dt'}|\psi_n(\lambda(t'))\rangle dt'. \tag{9.27}$$

The first term on the left gives the usual dynamical phase factor $e^{(-i/\hbar)\int E(t)dt}$ familiar from every introductory quantum mechanics text. The second term, which has nothing to do with the dynamics induced by the Hamiltonian is the **Berry phase**,

$$\gamma_n(C) = i\int_0^t \langle\psi_n(\lambda(t'))|\frac{d}{dt'}|\psi_n(\lambda(t'))\rangle dt'. \tag{9.28}$$

Here C is the path in parameter space. Making a change of integration variable and using the chain rule, it is clear that

$$dt'\frac{d}{dt'} = dt\,\frac{d\boldsymbol{\lambda}}{dt}\cdot\nabla_\lambda = d\boldsymbol{\lambda}\cdot\nabla_\lambda, \tag{9.29}$$

so the Berry phase can be written in the alternative form

$$\gamma_n(C) = i\int_0^t \langle\psi_n(\lambda(t'))|\nabla_\lambda|\psi_n(\lambda(t'))\rangle\cdot d\boldsymbol{\lambda}. \tag{9.30}$$

In particular, we are concerned with the case in which the parameters undergo cyclic evolution, returning to their initial values at the end of path C:

$$\gamma_n C = i\oint_C \langle\psi_n(\lambda(t'))|\nabla_\lambda|\psi_n(\lambda(t'))\rangle\cdot d\boldsymbol{\lambda}, \tag{9.31}$$

which generalizes equation (9.20) to continuous systems. The latter can be put into a more suggestive form,

$$\gamma_n = \oint_C \boldsymbol{A}_n(\boldsymbol{\lambda})\cdot d\boldsymbol{\lambda}, \tag{9.32}$$

by defining

$$\boldsymbol{A}_n(\lambda) = i\langle\psi_n(\lambda(t'))|\nabla_\lambda|\psi_n(\lambda(t'))\rangle. \tag{9.33}$$

Equation (9.32) is of the same form as the phase change produced by having a charged particle traverse a closed path through a gauge field (chapter 2), with $\boldsymbol{A}_n(\boldsymbol{\lambda})$ playing the role of the gauge potential or gauge connection. (However, the path in this case exists in some parameter space, not necessarily in real space.) So $\boldsymbol{A}_n(\boldsymbol{\lambda})$ is called the **Berry connection**. The analogy with gauge theory can be taken further: we may define a **Berry curvature**,

$$\mathcal{F}_{\mu\nu} = \partial_\mu A_{n,\nu} - \partial_\nu A_{n,\mu}, \tag{9.34}$$

that plays the same role as the electromagnetic field tensor (compare to chapter 2 and section 4.6). In terms of $\mathcal{F}_{\mu\nu}$, the Berry phase is

$$\gamma_n = \oint_S \mathcal{F}_{\mu\nu}dx^\mu\wedge dx^\nu, \tag{9.35}$$

where S is the area enclosed by curve C. $\mathcal{F}_{\mu\nu}$ is gauge invariant, so this latter form guarantees that the Berry phase around a closed loop is also gauge invariant.

The integral of curvature over an area is proportional to a solid angle, so the formula for the Pancharatnam phase as a solid angle (equation (9.9)) can now be seen as a special case of equations (9.34) and (9.35). The Aharonov–Bohm effect (see chapters 1 and 2) is also a special case, in which the solid angle is proportional to the quantized magnetic flux; in this case the phase is truly of topological, rather than strictly geometric, origin, with the quantum number labeling the homotopy class of path through the gauge field.

9.3 Geometric phase in optical fibers

We mention here one other example of geometric phases in optics. Soon after the publication of Berry's original work on geometric phases from adiabatic closed loops, a fiber optical version was proposed by Chaio and Wu [17] (who were apparently unaware at the time of Pancharatnam's work), and was demonstrated experimentally in [18]. An alternative derivation of this effect was given in [19], stressing its connection to the Gauss–Bonnet theorem.

Here, the fiber is wound into a helix, and linearly polarized light is sent into one end of the fiber. As the light propagates along the fiber, the wavevector $\boldsymbol{k}$ of the light traces out some path in momentum space. It is arranged so that the initial and final ends of the fiber are parallel, so that the momentum space path carried out by the light forms a closed loop, $\mathcal{C}$. Standard optical fibers are made of silica, in which nonlinear effects are negligible at normal intensities, so the magnitude of $\boldsymbol{k}$ remains constant and the path $\boldsymbol{k}$ traverses can be viewed as being confined to a sphere in momentum space. The closed loop $\mathcal{C}$ on the sphere will subtend a solid angle

$$\Omega(\mathcal{C}) = 2\pi N(1 - \cos\theta), \tag{9.36}$$

where N is the number of loops in fiber optical coil and θ is the pitch angle of the helix. The geometric phase is then again proportional to the solid angle,

$$\gamma(\mathcal{C}) = -\sigma\Omega(\mathcal{C}), \tag{9.37}$$

where $\sigma = \pm 1$ is the helicity of the light. Furthermore, the polarization direction of the light exiting at the end is rotated by an angle $\omega(\mathcal{C})$ from the polarization of the light that entered at the initial end.

9.4 Holonomy interpretation

The geometric meaning of these new phases should now be easy to see. At each point a Jones vector or a wavefunction can be multiplied by a phase factor, $e^{i\theta}$. Similar to the discussion of the local gauge principle in chapter 2 the value of θ can be varied independently at each point if we introduce a gauge field (in this case the Berry connection) with appropriate transformation properties to cancel its effect. The phases form a circle in the complex plane, and this circle is a representation of the one-dimensional $U(1)$ group. So we have a principle $U(1)$ bundle (see chapter 4) over the configuration space, where the circular fiber attached to each point represents the phase factor. The Berry connection and Berry curvature are the connection and curvature of this principle bundle. Now, if we periodically vary the parameters of the system (for example, the polarization in the Pancharatnam case or the wavevector direction in the case of the helical optical fiber), the system traces out a closed curve in the configuration space, which gets lifted to a curve in the bundle. However, this lifted curve is not necessarily closed: it may end at a different point in the circular fiber than the point where it started. The geometric phase is then simply the angle through which the final state must rotate to make the path close on the bundle. In other words, the Berry phase is an example of holonomy on a $U(1)$ bundle. Here, the energy levels were assumed to be nondegenerate, so the fiber is one-dimensional. In more general situations, the energy may be degenerate and so the states at that energy form a space of dimension greater than one, and the fiber will be more complicated. The bundle will still be a principle bundle, but the structure group will be non-Abelian [3]. The essential geometric picture, however, remains qualitatively the same.

References

[1] Berry M V 1984 *Proc. Roy. Soc. London* A **392** 45
[2] Pancharatnam S 1956 *Proc. Indian Acad. Sci.* A **44** 247
[3] Wilczek F and Zee A 1984 *Phys. Rev. Lett.* **52** 2111
[4] Aharonov Y and Anandan J 1987 *Phys. Rev. Lett.* **58** 1593
[5] Samuel J and Bhandari R 1988 *Phys.Rev. Lett.* **60** 2339
[6] Uhlmann A 1986 *Rep. Math. Phys.* **24** 229
[7] Zak J 1989 *Phys. Rev. Lett.* **62** 2747
[8] Hannay J H 1985 *J. Phys. A Math. Gen.* **18** 221
[9] Simon B 1983 *Phys. Rev. Lett.* **51** 2167
[10] Anandan J, Christian J and Wanelik K 1997 *Am. J. Phys.* **65** 180
[11] Nitayananda R 1994 *Curr. Sci.* **67** 238
[12] Hopf H 1931 *Math. Ann.* **104** 637
[13] Morandi G 1992 *The Role of Topology in Classical and Quantum Physics* (Berlin: Springer)
[14] Griffiths D F and Schroeter D F 2018 *Introduction to Quantum Mechanics* 3rd edn (Cambridge: Cambridge University Press)
[15] Shankar R 1994 *Principles of Quantum Mechanics* 2nd edn (Berlin: Springer)
[16] Kato T 1950 *J. Phys. Soc. Jap.* **5** 435
[17] Chaio R Y and Wu Y S 1986 *Phys. Rev. Lett.* **57** 933
[18] Tomita A and Chaio R Y 1986 *Phys. Rev. Lett.* **57** 937
[19] Ryder L H 1991 *Eur. J. Phys.* **12** 15

IOP Publishing

Topology in Optics (Second Edition)
Tying light in knots
David S Simon

Chapter 10

Topological states of matter

In chapter 1, the role of Dirac monopoles and the Aharonov–Bohm effect in spurring interest in topological aspects of quantum field theories was discussed. A similar situation occurred in condensed matter physics, where the unexpected discovery of the quantum Hall effect soon led to the revelation of a variety of related phenomena whose explanations also rely on topological effects. Among these are the discovery of materials that have quantized conductivities and electron states that propagate uni-directionally without scattering at impurities. Such states exist only on the edges of materials or on the boundaries between regions of the material that have different values of integer-valued topological invariants such as winding number or Chern numbers.

More recently, it has been realized that similar phenomena can be achieved in optical and photonic systems and that optical systems can in fact be used to simulate nontrivial topological behavior of condensed matter and other systems.

We begin in this chapter by reviewing the quantum Hall effect and other topological effects in condensed matter systems. (More extensive reviews of topological effects connected to the electronic band structure of solids include [1–5].) Then, in the next chapter, we describe how similar topological phenomena can arise in connection with motions of photons through optical systems.

10.1 The quantum Hall effect

The **classical Hall effect**, in which the presence of a static *magnetic* field causes the appearance of an *electric* potential difference, was discovered in 1879 by Edwin Hall while he was a graduate student at Johns Hopkins. The Hall effect forms the basis for numerous types of high-precision measurements and sensing devices.

Consider a conductor carrying a current along the x-axis in the presence of a uniform magnetic field pointing along the z-axis, as in figure 10.1. Let w and L be the widths in the y and z directions, respectively. Suppose the current along the x-axis is the result of a voltage difference V_0 between the ends of the material. The charge

doi:10.1088/978-0-7503-3471-6ch10

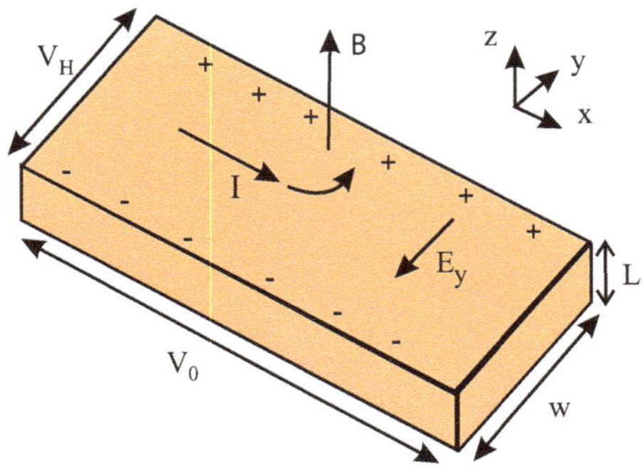

Figure 10.1. The classical Hall effect. The magnetic field deflects the current, causing a buildup of charges on the sides of the material. The separated charges then cause an electric field and a voltage difference in the direction transverse to the current.

carriers (electrons) move opposite to the current direction and feel a magnetic force $\boldsymbol{F}_B = -e\boldsymbol{v} \times \boldsymbol{B}$ in the y-direction. As a result, electrons start to accumulate on one side of the material, while a deficit of electrons creates an effective buildup of positive charge on the other side. The charge separation causes an electric field $\boldsymbol{E}_y$ to grow, pointing toward the left (from the positive charge toward the negative). This field then causes an electric force $\boldsymbol{F}_E = -e\boldsymbol{E}$ opposed to $\boldsymbol{F}_B$. The charge on the edges of the material continues to accumulate until it is sufficiently large for the electric and magnetic forces to cancel:

$$-eE_y = -e|\boldsymbol{v} \times \boldsymbol{B}| \quad \rightarrow \quad E_y = vB. \tag{10.1}$$

At this point, equilibrium is reached, with the charges on the edges of the conductor creating a voltage difference across the material in the y-direction,

$$V_H = E_y w = vBw. \tag{10.2}$$

This voltage is known as the **Hall voltage** and is easily measured by connecting a voltmeter across the two sides of the conductor.

The Hall voltage can be put into a more useful form. The current density J (defined as current per *length* in two dimensions) and the current I can be written in terms of the electron density (number of electrons per area) n as $J = nev$ and $I = Jw = nevw$. So the Hall voltage can be written in terms of the current as

$$V_H = \frac{BI}{ne}. \tag{10.3}$$

In addition to the usual Ohm's law resistance $R_{xx} = V_0/I$ (in this context referred to as the **longitudinal resistance**), we may also define a **transverse resistance**

$$R_{xy} = \frac{V_H}{I} = \frac{B}{ne}. \tag{10.4}$$

This resistance may seem a bit odd at first: it is measuring the ratio of the voltage difference in the y-direction to the current in the x-direction. But it turns out to be useful for making precision measurements. Since the experimenter usually controls the value of B, the measurement of the transverse Hall resistance provides an accurate way of determining the electron density n in the material. Alternatively, if the charge carrier density of the material is well known, then the Hall effect provides a means of making precise measurements of the magnetic field.

Suppose we set V_0 to zero and consider the behavior of individual electrons in the 2D system under the influence of an external magnetic field. The magnetic field causes the electrons to undergo circular motion (figure 10.2) in the xy-plane at the cyclotron frequency,

$$\omega = \frac{eB}{m} \tag{10.5}$$

Note, however, what happens at the edges: the circle cannot be completed without running into the edge and reflecting. This leads to the skipping motion shown in figure 10.2. The result is that the bulk of the conduction occurs only at the boundaries, and current conducts in only one direction at each boundary. These unidirectional boundary currents are said to be *chiral*.

The current density in two dimensions, the current per length, is $J = I/w$. The conductivity of the material relates the current density to the electric field,

$$\boldsymbol{J} = \sigma \boldsymbol{E}, \tag{10.6}$$

where the conductivity σ is a tensor described by a two-by-two matrix:

$$\sigma = \begin{pmatrix} \sigma_{xx} & \sigma_{xy} \\ -\sigma_{xy} & \sigma_{xx} \end{pmatrix}. \tag{10.7}$$

The resistivity is given by the inverse of the conductivity:

$$\rho = \sigma^{-1} = \begin{pmatrix} \rho_{xx} & \rho_{xy} \\ -\rho_{xy} & \rho_{xx} \end{pmatrix}. \tag{10.8}$$

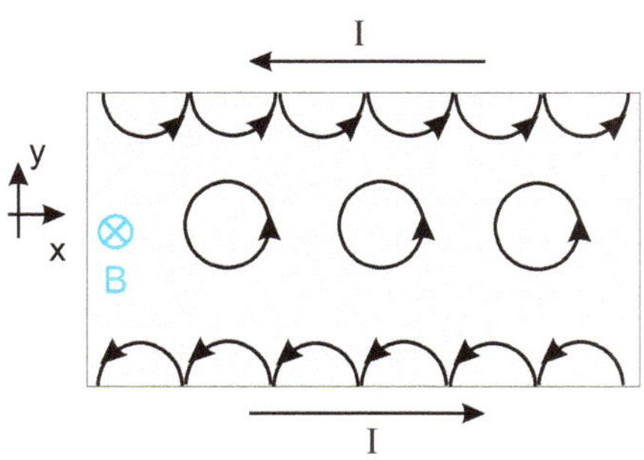

Figure 10.2. Motion of electrons in a magnetic field. In the bulk, the magnetic force causes circular motions. Near the edges, the circular motions are interrupted by collisions with boundary. This leads to unidirectional skipping motion along each edge. Note that the motion is in opposite directions on each of the two edges.

The matrices are easy to invert, giving

$$\sigma_{xx} = \frac{\rho_{xx}}{\rho_{xx}^2 + \rho_{xy}^2}, \qquad \sigma_{xy} = -\frac{\rho_{xy}}{\rho_{xx}^2 + \rho_{xy}^2}. \tag{10.9}$$

It should be noted that in the two-dimensional context, there is no distinction between resistance and resistivity; for example, since $V_H = -wE_y$ and $I = wJ_x$, it is seen that

$$R_{xy} = \frac{V_H}{I} = \frac{E_y}{J_x} = -\rho_{xy}. \tag{10.10}$$

The **Hall coefficient** (in m^3/C) is then defined as

$$R_H = -\frac{E_y}{J_x B} = \frac{\rho_{xy}}{B} = \frac{1}{ne}. \tag{10.11}$$

It is straightforward to compute the various quantities just defined by means of the **Drude** or **free-electron model** [6–8] of solid-state physics. In particular, the model predicts that:

$$\rho_{xx} = \frac{m}{ne^2\tau}, \qquad \rho_{xy} = \frac{B}{ne}, \tag{10.12}$$

where m, n, and τ are, respectively, the electron mass, electron density, and scattering time in the material, which again gives $R_H = 1/ne$. Notice in particular that in this classical model, the transverse resistivity is a linear function of magnetic field B.

The **integer quantum Hall effect** was discovered in 1980 [9]. The effect occurs in thin films (of thickness no more than 5–10 nm) and other systems where the electrons have little freedom to move in one direction, leaving an effective two-dimensional system. In the presence of a magnetic field perpendicular to the 2D system, electrons exhibit cyclotron motion, as above. At room temperature, these motions are disrupted by thermal noise, but at very low temperatures ($< 1\,K$) and strong magnetic fields the thermal scattering time is much longer than the orbital period, leaving the circular motions largely undisturbed. The energies of the cyclotron orbits are quantized, leading to a series of discrete energy levels $E_k = (k + 1/2)\hbar\omega$, called **Landau levels** (with $k = 1, 2, \ldots$). Under these conditions, the transverse resistivity ρ_{xy} exhibits a series of plateaus as a function of magnetic field strength (figure 10.3), in contrast to the linear dependence on B that we saw for the classical effect above. The plateaus appear at the values

$$\rho_{xy} = \frac{h}{\nu e^2} \approx \frac{25,812.807\,45\,\Omega}{\nu}, \tag{10.13}$$

where $\nu = 2, 3, 4, \ldots$. While on these plateaus, the ordinary (non-Hall) resistance of the material vanishes. The transverse conductivity is found to be quantized:

$$\sigma_{xy} = \frac{e^2}{h}\nu. \tag{10.14}$$

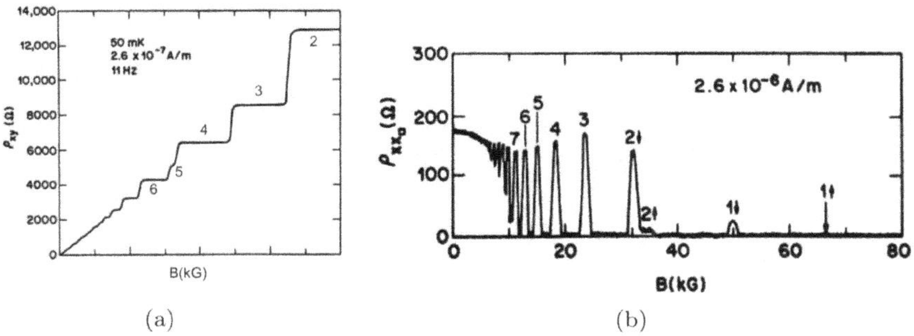

Figure 10.3. The integer quantum Hall effect. (a) As the magnetic field is increased, the transverse Hall resistance exhibits plateaus. Each plateau is characterized by an integer value of $h/R_H e^2$ or $\sigma h/e^2$. (b) The longitudinal resistivity along the direction of the E field vanishes on the plateaus, spiking at the points where the jumps between plateaus occur. (Figures reproduced with permission from [10], copyright 1982 American Physical Society, with additional labels added to the plateaus.)

The plateaus occur because the Landau levels are quantized: increasing B cannot affect ρ_{xy} until it contributes enough energy to promote an electron to the next higher Landau level. The number ν, called the **filling fraction**, measures the number of electrons per flux quantum.

Each plateau is centered at the field value

$$B = \frac{2\pi\hbar n}{\nu e} \equiv \frac{n}{\nu}\Phi_0, \tag{10.15}$$

where n is the electron density and $\Phi_0 = 2\pi\hbar/e$ is the fundamental flux quantum. The presence of the integer ν is a reflection of the topological origin of the integer quantum Hall effect. On each plateau, the longitudinal resistivity ρ_{xx} vanishes, while $\rho_{xy} \neq 0$. This leads to an odd effect: according to equation (10.9), this implies that σ_{xx} also vanishes: the longitudinal resistivity and conductivity both vanish simultaneously!

The integer quantum Hall effect is prominent in impure or disordered material samples. As the sample becomes purer, the integer plateaus begin to shrink, and eventually disappear. As this happens, a new effect becomes apparent: plateaus at fractional values begin to become noticeable. This **fractional quantum Hall effect** was discovered in semiconductor materials in 1982 [11, 12]. The plateaus again occur at $\rho_{xy} = h/e^2\nu$, but now ν takes on simple fractional values such as $\nu = 1/3, 1/5, 1/7, 2/3, 2/5, \ldots$. In contrast to the integer effect, which occurs even if the interactions between electrons are ignored, the fractional effect is dependent on the presence of electron–electron interactions.

The transverse Hall conductivity is of topological origin. At each point in the two-dimensional Brillouin zone, we can define

$$F = \partial_{k_x} A_y - \partial_{k_y} A_x, \tag{10.16}$$

which is the curvature of a Berry potential $\boldsymbol{A}$. The Brillouin zone is a torus, T^2, in momentum space, so the value of F averaged over the zone is

$$\sigma_{xy} = \frac{1}{(2\pi)^2}\int_0^{2\pi}\int_0^{2\pi}\sigma_{xy}(\boldsymbol{k})dk_x dk_y = \frac{e^2}{(2\pi)h}\int_{T^2} F\, d^2k = \frac{e^2}{h}c_1, \qquad (10.17)$$

where c_1 is the first Chern number defined in equation (5.14) and the constants were inserted to give units of conductance. In other words, the integer in equation (10.13) is a Chern number. In the case where multiple energy bands are occupied, the integer is given by the sum of the Chern numbers of all occupied bands. The inability to continuously vary between the integer-valued topological invariants is what leads to the quantization of the transverse conductivity. In this context, the sum of the Chern numbers of the occupied bands on the Brillouin zone is referred to as the **TKNN invariant** [13].

Further odd phenomena occur in the fractional quantum Hall effect. For example, the charge carriers are 'fractionalized': screening of charges leads to the electrons seeming to have parts of their charge in different, widely separated locations. These charges are attached to multi-particle states formed from many mutually interacting electrons, or in other words to quasiparticle excitations. The quantum mechanical statistics of these excitations also become fractional. In three dimensions all particles must have integer or half-integer spin, leading to bosonic or fermionic behavior, respectively. But in two dimensions, a broader range of possible statistical behaviors can arise, interpolating between bosonic and fermionic behavior. Such fractional-statistics particles are called **anyons** and will come up again briefly in the next chapter.

The relevance of the Chern number here is due to the following. The Hamiltonian is a matrix-valued function on the two-dimensional Brillouin zone parameterized by the two components of the momentum, k_x and k_y. Because of the periodicity of the lattice, these components are both periodic; thus, the Brillouin zone is a torus, T^2. Meanwhile, the Hamiltonian of the system can be written as a matrix similar to that of the SSH model (see the next section), representing states lying on a Bloch sphere state, S^2. There is thus a map from the torus to the sphere. These are classified by the homotopy groups of bundles of states over tori, which are indexed by an integer, the Chern number (or equivalently the Pontrjagin number).

Topological effects can occur in one-dimensional systems as well. In this case, the relevant topological invariant becomes the winding number. We'll see this explicitly for a simple example in section 10.2.

Other variations on the quantum Hall effect can occur. In 1988, Haldane [14] showed that a Hall-like effect can occur in periodic systems in the absence of Landau levels, providing a quantum version of an effect first seen in classical systems by Hall in 1880. This **anomalous quantum Hall effect** was seen experimentally in 2013 [15]. As another example, note that the quantum Hall effect can be viewed as a unidirectional charge pump: charge is transported in a single direction, with no charge flowing the other way; a similar effect, known as the **spin quantum Hall effect**, acts as a one-way pump for spin [16–18]. Spin up electrons travel in one direction, while spin down travels the other way.

Materials that exhibit the quantum Hall effect are examples of a more general category of materials called **topological insulators**, materials that are insulating in their bulk but conduct unidirectionally on their boundaries. The surface currents are immune to backscattering at impurities, and so the currents are very robust. These materials are currently objects of intense interest, both because of their unique physical properties and because of their applications for quantum information processing and other areas.

More detailed reviews of quantum Hall effects may be found in [19–22]. Reviews of topological insulators the related topic of topological superconductors include [4, 23–25].

Before discussing the role of topology in more complex material systems, we first discuss a simpler one-dimensional system in the next section.

10.2 One-dimensional example: the SSH model

To clarify further how the topological considerations arise in a material system, it is helpful to look at one of the simplest systems in which they arise, the **Su–Schreiffer–Heeger (SSH) model** [25–28]. The model was proposed to describe the electronic properties of polyacetylene chains, but a mathematically equivalent structure was also discovered in high-energy physics by Jackiw and Rebbi while studying fermionic solitons [29].

Consider a one-dimensional lattice of length $2N$. To avoid edge effects for now, we can either take $N \to \infty$ or we can use periodic boundary conditions, connecting the two ends of the lattice together to form a loop. We further assume that the lattice is made up of two alternating types of objects, labeled A and B, with each unit cell of the lattice therefore being made up of two lattices sites, one of each type (figure 10.4). The unit cells are labeled by an integer, $m = 1, 2, \ldots, N$. But v, the intracell hopping amplitude (the amplitude to jump between A and B within the same cell, leaving m unchanged) can differ from the intercell hopping amplitude, w (hopping between different unit cells, leading to $\Delta m = \pm 1$). Often A and B will represent single-valence-electron atoms; recalling that each energy level in an atom can hold two electrons in opposite spin states, that means that the valence band is half full. There are therefore energetically available unoccupied states around for the electrons to hop into.

The system can then be described by the Hamiltonian

$$
\begin{aligned}
H = \sum_m \{ & v(|m, B\rangle\langle m, A| + |m, A\rangle\langle m, B|) \\
& + w(|m+1, A\rangle\langle m, B| + |m, B\rangle\langle m-1, A|)\}.
\end{aligned}
\tag{10.18}
$$

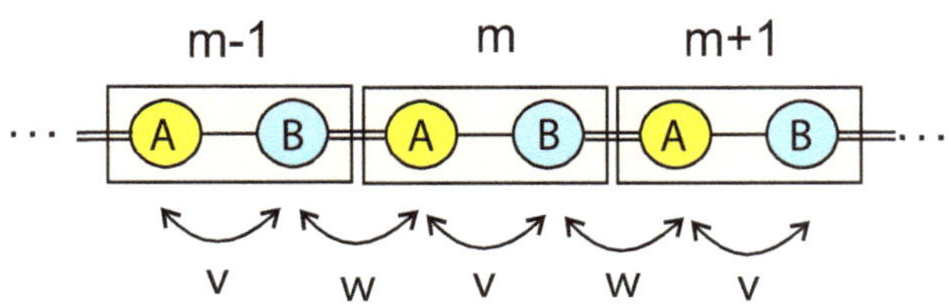

Figure 10.4. The SSH model. A lattice is made up of two interspersed sublattices (A and B), with an adjacent pair of A and B sites making up a unit cell. The locations of the unit cells are labeled by an integer m. Transitions within a cell can have a different amplitude (v) than transitions between cells (w).

Each of the four terms represents one of the four hopping possibilities: either left or right, and either within the same cell or between two cells. In matrix form, the Hamiltonian can be expressed as

$$H = \begin{pmatrix} 0 & v & 0 & 0 & 0 & 0 & \dots \\ v & 0 & w & 0 & 0 & 0 & \\ 0 & w & 0 & v & 0 & 0 & \\ 0 & 0 & v & 0 & w & 0 & \\ 0 & 0 & 0 & w & 0 & v & \\ \vdots & & & & & & \ddots \end{pmatrix} \tag{10.19}$$

We think of 'residing at A' and 'residing at B' to be two separate states that exist at each unit cell and treat A and B as two values of some 'internal' variable of the particle, similar to spin or polarization. The state at each unit cell can therefore be described by a two component vector, with the upper entry representing the amplitude of being found in state A and the lower entry being the amplitude of being in state B. As a result, we can think of the A/B part of the Hamiltonian as describing a two-state system like electron spin and write the Hamiltonian in terms of Pauli matrices (see equation (10.24)). But first we perform a Fourier transform from unit cell position m to a dimensionless momentum-like variable k. Define

$$|k\rangle = \frac{1}{\sqrt{N}} \sum_{m=1}^{N} e^{imk} |m\rangle, \tag{10.20}$$

where the allowed values of k are $k = \{2\pi/N, 4\pi/N, \dots 2\pi\}$.

Henceforth, we will assume that N is large so that we can approximate k as a continuous variable ranging from 0 to 2π. Notice that the state $|k\rangle$ is periodic under $k \to k + 2\pi$. So we can always restrict ourselves to the range $0 < k < 2\pi$, the fundamental Brillouin zone of the one-dimensional system. Because of the periodicity, we can think of the two ends of the Brillouin zone wrapping around to form a circle S^1. k is called the **quasimomentum** or **crystal momentum**.

As a result of taking the Fourier transform to k-space, the Hamiltonian block diagonalizes, with a two-by-two block for each fixed k. Some algebra leads to blocks of the form

$$H(k) = \begin{pmatrix} 0 & v + w\, e^{-ik} \\ v + w\, e^{+ik} & 0 \end{pmatrix}. \tag{10.21}$$

Since this is a two-state system for each k, we expect there to be two energy levels. These are easily found by diagonalizing the k-space Hamiltonian. Putting normalization aside, the eigenvectors are given by

$$|\psi_\pm(k)\rangle = \begin{pmatrix} \pm\, e^{-i\phi(k)} \\ 1 \end{pmatrix}, \tag{10.22}$$

where $\tan\phi(k) = w \sin k / v + w \cos k$, and the energy eigenvalues are

$$E(k) = \pm\sqrt{v^2 + w^2 + 2vw \cos k}\,. \tag{10.23}$$

(Note that eigenstates are states of definite k; their Fourier transforms are therefore completely delocalized in position space, with equal amplitude of being found at any unit cell.)

These energies are plotted in figure 10.5 for several ranges of v and w values. Several things should be noticed about these plots. First, the two energies of the system come in opposite sign pairs; this is a consequence of symmetry as we will see later. Second, there is a gap between the two energy levels as long as $v \neq w$. Thus, for $v \neq w$ we can think of the system as an insulator: an electron in the 'valence band' (the lower energy state) cannot jump to the 'conduction band' (the higher state) without a finite energy input to the system. However, the case $v = w$ is anomalous: under this condition the energy gap closes and the system becomes an insulator.

To see where topology comes in, we start by writing the Hamiltonian in a different form. Any 2×2 Hamiltonian can be written in terms of the Pauli matrices and the identity:

$$H(k) = h_0(k)I + h_x(k)\sigma_x + h_y(k)\sigma_y + h_z(k)\sigma_z, \tag{10.24}$$

where

$$\begin{aligned} I &= \begin{pmatrix} 1 & 0 \\ 0 & 1 \end{pmatrix}, \quad \sigma_x = \begin{pmatrix} 0 & 1 \\ 1 & 0 \end{pmatrix}, \\ \sigma_y &= \begin{pmatrix} 0 & -i \\ i & 0 \end{pmatrix}, \quad \sigma_z = \begin{pmatrix} 1 & 0 \\ 0 & -1 \end{pmatrix}. \end{aligned} \tag{10.25}$$

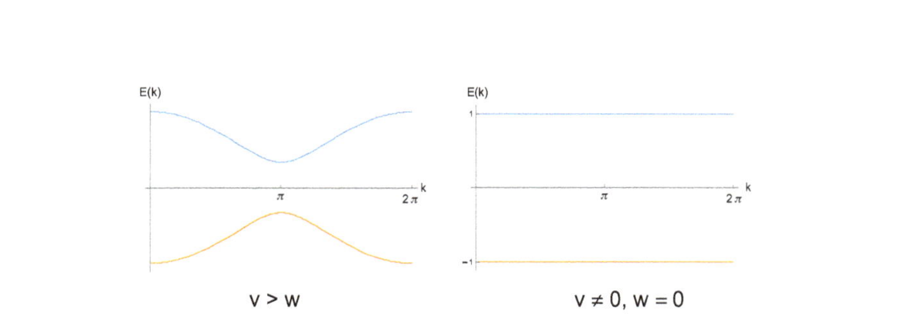

Figure 10.5. An energy gap exists between the levels of the SSH system, indicating that the system is insulating. However, when the hopping parameters obey $v = w$, the gap closes. At this point, the winding number becomes undefined and transitions between states of different winding number become possible.

The h_0 term is simply a constant that can always be removed by an overall global shift of all energies. For the SSH model, the remaining coefficients are

$$h_x(k) = v + w \cos k, \qquad h_y(k) = w \sin k, \qquad h_z = 0. \tag{10.26}$$

So if we think of $H(k)$ as defining a three-dimensional vector with components $\{h_x, h_y, h_z\}$, this vector is actually confined to a two-dimensional plane. In fact, as k varies from 0 to 2π, it traces out a circle in the xy plane (figure 10.6).

The case $v = w$ at which the energy gap closes is a singular point of the system, at which a change of topology occurs. This can be seen in figure 10.6. In one dimension, the Brillouin zone is a circle. The map $k \to \boldsymbol{h}(k)$ therefore defines an element of $\pi_1(S^1)$, characterized by a winding number ν. For $v = w$ (the central image in figure 10.6) the image of this map is tangent to the origin, so the winding number around the gapless point $\boldsymbol{h}(k) = 0$ is undefined. For $v < w$ (left-hand figure), the winding number is $\nu = 1$, while for $v > w$ (right-hand figure) it is $\nu = 0$.

As k varies across the Brillouin zone, the energy eigenstates pick up a nonzero Berry phase (or more specifically, a Zak phase), which can vary by either 0 or π around the closed loop. So the map $k \to e^{i\phi(k)}$ has a winding number of either 0 or 1: the topological state of the system is characterized by a $\mathbb{Z}_2$ invariant, taking values 0 or 1. If two SSH chains with different topological invariant are joined together, a localized, topologically protected zero-energy edge state will arise at the boundary. This state interpolates between different topological quantum numbers on the two sides of the boundary, acting as a topological soliton, but it cannot propagate in either of the bulk regions, which forces it to remain at the boundary. The energy gap has to close at the boundary, providing a point at which the winding number is not well-defined.

Notice that if an SSH chain terminates at a boundary, the empty space beyond the end can be thought of as a trivial (zero-winding number) insulator; therefore topologically protected boundary states should exist at the edges of the system as well.

In realistic solids, the isolated energy *levels* of the SSH system become energy *bands*. Going from this simple one-dimensional model to real two-dimensional systems, many of the qualitative properties described above are preserved, but with

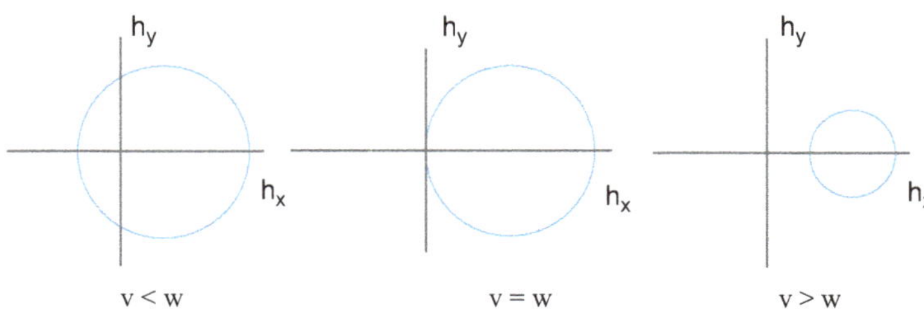

Figure 10.6. As k varies over the Brillouin zone, the vector $h_x(k)$, $h_y(k)$ traces out a circle in the plane. For $v = w$ the circle is tangent to the origin, so the winding number around the gapless point $\boldsymbol{h}(k) = 0$ is undefined. For $v < w$, the winding number is $\nu = 1$, while for $v > w$ it is $\nu = 0$. So the parameter values $v = w$ define a topological transition point.

new features appearing. For example, the edge of the system will now be one-dimensional, so that the boundary states can propagate along it (recall the unidirectional states that skip along the boundary of the quantum Hall system). The Brillouin zone is now a torus, and the appropriate topological invariant becomes the first Chern number or, more generally, the TKNN invariant. Each band has its own Chern number and the TKNN invariant is formed by summing them over the occupied bands.

10.3 Topological phases and localized boundary states

As the quasi-momentum of a crystal or other periodic system is varied over a full Brillouin zone, the momentum–space Hamiltonian $\hat{H}(\boldsymbol{k})$ traces out a closed loop in the space of Hermitian matrices. The winding number counts the number of times the loop encloses the singular point of the Hamiltonian at the origin, where all of the energies collapse to zero. In the case of a two-dimensional system, the values of k will lie on a torus and $\hat{H}(\boldsymbol{k})$ will define a mapping of the torus to the Bloch sphere; the topological invariant associated with this mapping will be a Chern number.

Topological insulators have energy levels or energy bands that are separated by finite gaps (figure 10.7(a)). The size of the gap varies as k traverses the Brillouin zone, but should remain nonzero to avoid the zero-energy singularity. Topological invariants can only change when the energy gap closes; the values of topological invariants can only change when the energy bands cross the singular point, where the invariant becomes undefined. The wavefunctions in each energy band form a fiber bundle over the Brillouin, and the topological invariants measure the topological nontriviality of the bundle.

Each band will have its own Chern number. Time-reversal invariance implies that the bands come in opposite Chern number pairs, so that the sum of the Chern numbers of all the bands vanishes: $\sum_j c_1^{(j)} = 0$, where j labels the bands. A nonzero

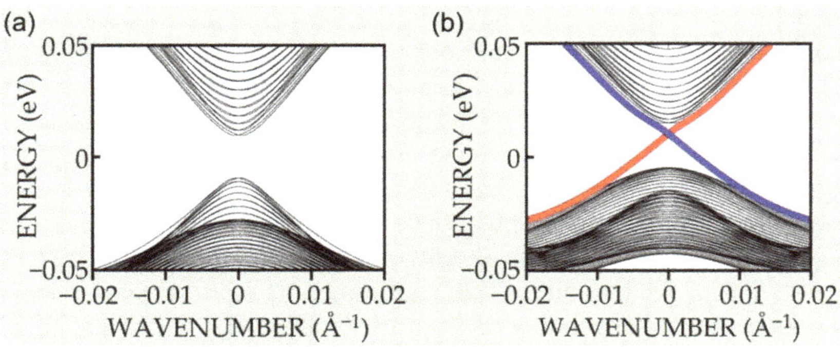

Figure 10.7. (a) In topological insulators, finite energy gaps are maintained between bands. Topological invariants such as the Chern number can only change when the gap closes. (b) When two materials with different Chern number are brought together, new states appear at the boundary that can cross between the bands. These states are highly localized and highly stable: topological considerations prevent them from being destroyed by continuous changes of the system parameters. (Figures reproduced with permission from [18], copyright 2009 American Institute Physics.)

value of $c_1^{(j)}$ indicates that there is a topological obstruction preventing wavefunctions of states in that band from being globally defined; instead, the wavefunctions need to be defined on a set of overlapping open sets, with transition functions relating the wavefunctions on the overlap regions. Chern numbers are only defined in even dimensions (in particular, the first Chern number is defined in two dimensions), but in other dimensions the Hopf invariant or other topological invariants play a similar role.

For the sake of specificity, let us assume henceforth that we are dealing with two dimensional systems; in other words, three-dimensional objects in which one direction is of negligible thickness. A common example is graphene, which forms two-dimensional layers that are only a single atom thick.

Suppose that a pair of two-dimensional crystals, M_1 and M_2, of different Chern number $c_1(M_1) \neq c_1(M_2)$, are brought into contact. Then it is found that a new quantum state arises, which is highly localized near the one-dimensional boundary in the sense that it decays rapidly as you move away in either direction. These **boundary or edge states** interpolate between the solutions on the two sides and only appear when the gap between energy bands vanishes. These zero-energy states are topological solitons, and they are of great interest for information processing and other applications because they are extremely robust: they are defined by the global properties of the system and cannot be destroyed by continuous perturbations of the system's Hamiltonian. The states are said to be **topologically protected**.

There are two types of boundary states: one whose energy crosses zero in the positive direction (going from the lower energy valence band to the higher energy conduction band as the electron moves from left to right) and those that cross zero in the negative direction (figure 10.7(a)). These unidirectional boundary states are said to have positive and negative chirality, respectively, and their group velocities are given by the slopes of their energy curves:

$$v = dE/dk. \tag{10.27}$$

Local disorder and imperfections may alter the slope of a given boundary state, but will not affect the fact that it must continue to slope in the same direction to maintain its band connections, which gives a qualitative explanation of the unidirectional nature of the boundary states. The example shown in figure 10.7(b) shows both leftward and rightward moving edge states; in the presence of time reversal symmetry both states must exist. When the time reversal symmetry is broken it is possible to have just a rightward moving state without a left-moving partner (or vice versa) on a given boundary. Without counterpropagating states available of the same energy, the boundary state becomes immune to backscattering or to scattering into the bulk, even in the presence of large amounts of disorder or impurity. When an imperfection occurs at the edge, the state simply extends a little farther into the bulk in order to bypass the disturbance and then returns back to the boundary to continue on its way.

Let $N_\pm$ represent the numbers of these two types of states. We can define a type of topological index for this system,

$$\nu = N_+ - N_-. \tag{10.28}$$

Then a fundamental result is the bulk-boundary correspondence:

$$\nu = c_1(M_2) - c_1(M_1). \tag{10.29}$$

In other words, the difference in the number of boundary states going in each direction must equal the change in the Chern number across the boundary:

$$N_+ - N_- = c_1(M_2) - c_1(M_1); \tag{10.30}$$

the values of the topological invariants in the bulk determine the number of topologically protected states localized on the boundary. Note that this is essentially an index theorem, very reminiscent of the Atiyah–Singer theorem: it relates topological invariants to the number of solutions of a differential equation.

10.4 The role of discrete symmetries

The type of topological invariant characterizing a system is determined primarily by the number of dimensions of the system and the symmetries that are present. Discrete symmetries are of particular importance.

A system is said to have a unitary symmetry if its Hamiltonian H is preserved by a unitary operator, U:

$$UHU^{\dagger} = H. \tag{10.31}$$

Such symmetries should be familiar from introductory quantum mechanics classes. Common examples include time-translation symmetry (a shift of the origin of the time coordinate) or spatial translation symmetry (a shift in the spatial origin). These are both continuous symmetries, and according to Noether's theorem each such continuous symmetry leads to a conservation law. The conservation laws for the two symmetries just mentioned are energy conservation and momentum conservation, respectively.

Discrete symmetries are also important in quantum mechanics. They are our main focus here because of their importance in classifying topological phases. In particular we will be interested in chiral symmetries, time reversal symmetry, and charge conjugation (or particle–hole) symmetry.

In addition to symmetries of the type obeying equation (10.31), it is possible to have discrete **chiral symmetries**, which obey

$$\Gamma H(\boldsymbol{k})\Gamma^{\dagger} = -H(\boldsymbol{k}). \tag{10.32}$$

Here, the operator Γ must be both Hermitian ($\Gamma^{\dagger} = \Gamma$) and unitary ($\Gamma^{\dagger} = \Gamma^{-1}$). These two relations imply that the chiral operator squares to the identity operator:

$$\Gamma^2 = \Gamma\Gamma^{-1} = 1. \tag{10.33}$$

The SSH model of the previous section has such a chiral symmetry. Define projection operators onto each of the two sublattices:

$$P_A = \sum_{m=1}^{N} |m, A\rangle\langle m, A| \quad \text{and} \quad P_B = \sum_{m=1}^{N} |m, B\rangle\langle m, B|. \tag{10.34}$$

These obey $P_A^2 = P_A$, $P_B^2 = P_B$, $P_AP_B = 0$, and $P_A + P_B = 1$. Then it can be verified that the operator

$$\Gamma = P_A - P_B \tag{10.35}$$

acts as a chiral symmetry on the SSH Hamiltonian. The effect of Γ is to reverse the sign of the wavefunction on one sublattice relative to the other sublattice. As a result, the chiral symmetry is also sometimes called a **sublattice symmetry**.

Chiral symmetries tell us some interesting things about the energy spectrum of a system. Suppose that $|\psi_n\rangle$ is an eigenstate of the Hamiltonian with eigenvalue E_n: $H|\psi_n\rangle = E_n|\psi_n\rangle$. Then using relation (10.32), it follows that the state $\Gamma|\psi_n\rangle$ has energy $-E_n$: $H\Gamma|\psi_n\rangle = -E_n\Gamma|\psi_n\rangle$. Thus, the nonzero energy eigenvalues of a chirally symmetric system come in opposite sign pairs, $\pm E_n$. It can also be shown that the eigenstates $|\psi_n\rangle$ are equally spread over the two sublattices when the energy is nonzero.

When the energy vanishes, $E_n = 0$, the situation is a bit different. In this case, the zero-energy eigenstate is self-conjugate, up to a sign, $\Gamma|\psi_n\rangle = \pm|\psi_n\rangle$. For the SSH model, these chiral eigenstates can be chosen so that each has support on just one sublattice.

Another important type of discrete symmetry is **time reversal symmetry** $\mathcal{T}$, which reverses the direction of time. It leaves position operators unchanged, but inverts momentum operators:

$$\mathcal{T}\hat{x}\mathcal{T}^{-1} = \hat{x}, \qquad \mathcal{T}\hat{\boldsymbol{k}}\mathcal{T}^{-1} = -\hat{\boldsymbol{k}}. \tag{10.36}$$

Since the canonical commutation relation between position and momentum, $[\hat{x}, \hat{\boldsymbol{k}}] = i\hbar$ must be preserved, this implies that $\mathcal{T}$ acts to conjugate complex numbers: $\mathcal{T}i\mathcal{T}^{-1} = -i$. Such an operator is referred to as **anti-unitary**. More generally, anti-unitary operators act on quantum states according to $\langle\mathcal{T}\phi|\mathcal{T}\psi\rangle = \langle\phi|\psi\rangle^*$ (where * is complex conjugation). This is in contrast to unitary operators which obey $\langle U\phi|U\psi\rangle = \langle\phi|\psi\rangle$. In general, the time reversal operator can always be decomposed into a unitary operator U and a complex conjugation operator K, $\mathcal{T} = UK$, where the form of U will depend on the type of system being considered. The action on the Hamiltonian is

$$\mathcal{T}H(\boldsymbol{k})\mathcal{T}^{-1} = H^*(-\boldsymbol{k}) \tag{10.37}$$

The time reversal operator must square to ± 1 times the identity operator: $\mathcal{T}^2 = \pm 1$. To be specific, $\mathcal{T}^2 = +1$ for spinless systems, while $\mathcal{T}^2 = -1$ for spin-1/2 systems. For spin-1/2 systems, time reversal symmetry has an important consequence known as **Kramer's degeneracy**: each nonzero energy level must be at least doubly degenerate, since $|\psi_n\rangle$ and $\mathcal{T}|\psi_n\rangle$ are orthogonal states with the same energy.

An additional type of discrete symmetry of importance in condensed matter systems is **particle–hole** or **charge conjugation symmetry**, $\mathcal{C}$, which interchanges particles and holes, and which obeys $\mathcal{C}^2 = \pm 1$. The action on the Hamiltonian is

$$\mathcal{C}H(\boldsymbol{k})\mathcal{C}^{-1} = -H^*(-\boldsymbol{k}) \tag{10.38}$$

If a system has both time reversal and charge conjugation invariance, then it automatically has a chiral symmetry defined by the operator

$$\Gamma = \mathcal{T} \cdot \mathcal{C}. \tag{10.39}$$

Given the discrete symmetries and the dimension of the system, it is possible to predict the topological classification of the system. It turns out that the topological invariant will take values in the set of integers $\mathbb{Z}$, the set of even integers $2\mathbb{Z}$, or the two-element set of integers mod 2, $\mathbb{Z}_2 = \{0, 1\}$. The result is the periodic table of topological insulators [30, 31] shown in table 10.1. For more detailed reviews, see [23, 32].

Floquet systems are those in which the Hamiltonian is periodically driven, for example by having the parameters of the system vary periodically in time. Often, if the particles being described are moving monotonically in some direction (the z-axis, say), then the z-direction is used as a surrogate for time and so the driving can be accomplished by a periodic spatial variation of the system; an example of this type of system will be seen in the next chapter. A periodic table similar to table 10.1 can also be constructed [33] for Floquet topological insulators, which shows a richer set of possible group structures. One characteristic feature of topological Floquet systems is that they are periodic in the energy as well as the time, since the phase factor $e^{-iET/\hbar}$ is unchanged when $E \to E + 2\pi\hbar/T$. As a result of this periodicity, the Floquet system can have two energy gaps, which (in dimensionless energy units) can always be taken to be at $E = 0$ and $E = \pi$. (The unit of energy here is taken to be $\hbar/T$, where T is the period.)

One further symmetry that will come up later is **parity** or **inversion symmetry**, which inverts spatial coordinate: $(x, y, z) \to (-z, -y, -z)$. This of course also

Table 10.1. Periodic table of topological insulators. The columns for the three symmetries $\mathcal{T}$, $\mathcal{C}$, Γ have entries 0 if the symmetry is absent, 1 if it is present (in the case of chiral symmetry Γ) or $\pm$ if it is present and squares to ± 1 (in the case of $\mathcal{T}$ and $\mathcal{C}$). The numbers in the top row indicate the dimension d of the material's bulk. In the columns under the dimensions, 0 means the system is topologically trivial; otherwise, the relevant topological invariant takes values in the integers ($\mathbb{Z}$), the integers mod 2 ($\mathbb{Z}_2 = \{0, 1\}$), or the even integers (denoted $2\mathbb{Z}$).

Class	$\mathcal{T}$	$\mathcal{C}$	Γ	0	1	2	3	4	5	6	7
A	0	0	0	$\mathbb{Z}$	0	$\mathbb{Z}$	0	$\mathbb{Z}$	0	$\mathbb{Z}$	0
AIII	0	0	1	0	$\mathbb{Z}$	0	$\mathbb{Z}$	0	$\mathbb{Z}$	0	$\mathbb{Z}$
AI	+	0	0	$\mathbb{Z}$	0	0	0	$2\mathbb{Z}$	0	$\mathbb{Z}_2$	$\mathbb{Z}_2$
BDI	+	+	1	$\mathbb{Z}_2$	$\mathbb{Z}$	0	0	0	$2\mathbb{Z}$	0	$\mathbb{Z}_2$
D	0	+	0	$\mathbb{Z}_2$	$\mathbb{Z}_2$	$\mathbb{Z}$	0	0	0	$2\mathbb{Z}$	0
DIII	—	+	1	0	$\mathbb{Z}_2$	$\mathbb{Z}_2$	$\mathbb{Z}$	0	0	0	$2\mathbb{Z}$
AII	—	0	0	$2\mathbb{Z}$	0	$\mathbb{Z}_2$	$\mathbb{Z}_2$	$\mathbb{Z}$	0	0	0
CII	—	—	1	0	$2\mathbb{Z}$	0	$\mathbb{Z}_2$	$\mathbb{Z}_2$	$\mathbb{Z}$	0	0
C	0	—	0	0	0	$2\mathbb{Z}$	0	$\mathbb{Z}_2$	$\mathbb{Z}_2$	$\mathbb{Z}$	0
CI	+	—	1	0	0	0	$2\mathbb{Z}$	0	$\mathbb{Z}_2$	$\mathbb{Z}_2$	$\mathbb{Z}$

inverts momentum: $(k_x, k_y, k_z) \rightarrow (-k_z, -k_y, -k_z)$. The corresponding operator $\mathcal{P}$ squares to +1 and acts on the momentum-space Hamiltonian as

$$\mathcal{P}H(\boldsymbol{k})\mathcal{P}^{-1} = H(-\boldsymbol{k}). \tag{10.40}$$

The presence of inversion symmetry or other spatial symmetries will alter the classification scheme for topological systems given above [32].

10.5 Varieties of topological insulators and related systems

The integer quantum Hall effect [9] was discovered experimentally in 1980, and its topological significance was soon recognized [13, 34]. The fractional quantum Hall effect [11] followed in 1983. Quantum Hall systems became the prototype for topological insulators, materials that insulate in the bulk but conduct on their surface and led to the idea of topological order [35]. But other variations soon followed.

The usual quantum Hall effect relies on the presence of a magnetic field to break time reversal symmetry, which allows unidirectional motion along the edges of the system. In 1988, Haldane [14] proposed a means of breaking time reversal without a net magnetic flux through the system, known as the anomalous quantum Hall effect; this anomalous effect was seen experimentally in 2013 [15].

It has also been found that instead of using either electrons or quasiparticles built exclusively from electrons to carry the surface currents, there are materials in which the carriers are quasiparticles formed from coherent superpositions of electrons and holes. These are known as **topological superconductors** [4, 23–25, 36], and the particle–hole excitations are strongly reminiscent of Cooper pairs in standard superconductors.

The **quantum spin Hall effect** (QSHE) [16–18] also occurs in systems where time reversal symmetry remains unbroken. On each boundary, spin-up states move in one direction, while spin-down states move the other way. Time reversal flips both the momentum and spin directions, maintaining the unidirectional flow of each spin state. As a result, each spin state remains highly immune to scattering into the opposite momentum direction. The QSHE takes advantage of strong spin–orbit couplings (interactions involving the spin and the quasimomentum) in materials such as graphene. Such materials are characterized by a $\mathbb{Z}_2$-valued invariant, known as the **Kane–Mele invariant** [16]. It was proposed [37] that the spin quantum Hall effect could be observed experimentally in HgTe quantum wells, and was later seen experimentally [38]. Up to now, what we have been discussing exist only in quasi-two-dimensional systems, but this $\mathbb{Z}_2$ topological phase can occur in three dimensions as well [39–41]

There are two-dimensional cases where topological effects occur despite vanishing Chern number. This happens for example when time reversal interchanges two so-called high symmetry points. In this case, there is an energy valley between the two symmetry points, and topological effects can occur in the valley, characterized by the **valley Chern number**, C_v [42–44]. Another situation in which distinct topological states can exist with vanishing Chern number is in the case of Floquet systems, where the Hamiltonian is a function of time [42, 43, 45, 46].

Additional variations include mirror-Chern [47–49] and higher-order topological insulators [50–53]. It has recently proposed that topological insulators on fractal lattice structures may also exist [54].

10.6 Dirac, Majorana, and Weyl points

It can happen that the energy gap between bands can close at isolated points in the Brillouin zone. When the band gap between the conduction and valence bands of a two-dimensional system vanishes at an isolated point, that point is called a **Dirac point**. (In three dimensions, when the bands touch at an isolated point the material is then referred to as a **semi-metal**.) Near the Dirac point, the Hamiltonian linearizes as a function of momentum k and looks like a two-dimensional analog of the four-dimensional Dirac Hamiltonian that describes spin-1/2 particles in high-energy physics. At this point, quasiparticle excitations appear that obey the dispersion relation (the energy–momentum relation) of a relativistic particle. Surfaces enclosing the Dirac point have quantized values of Berry flux (the integral of the Berry curvature). Each Dirac point is stable under local perturbations. They always arise in pairs as a result of the Nielsen–Ninomiya fermion-doubling theorem [55], and can only disappear by annihilating in pairs.

Normally, isolated gapless points between the bands are highly unstable. Small perturbations of the Hamiltonian (due to impurities in the material or thermal fluctuations, for example) are usually sufficient to destroy them. However, they can be stabilized by symmetries. When the Hamiltonian has both inversion and time reversal symmetry (PT) symmetry, pairs of stable Dirac points are topologically protected. As long as both symmetries hold, the Dirac point remains gapless (left side of figure 10.8). In the vicinity of each point, the Hamiltonian has the linear form $H(\boldsymbol{q}) = \hbar v(q_x\sigma_x + q_y\sigma_z)$, where $\boldsymbol{q}$ is the displacement from the Dirac point K, $\boldsymbol{q} = \boldsymbol{k} - \boldsymbol{K}$ and v is the group velocity. This Hamiltonian does not describe the behavior of a single electron, but rather of a coherent many-particle superposition.

The actions of the P and T symmetries on the Hamiltonian above are:

$$\mathcal{P}H(\boldsymbol{q})\mathcal{P}^{-1} = H(-\boldsymbol{q}) \tag{10.41}$$

$$\mathcal{T}H(\boldsymbol{q})\mathcal{T}^{-1} = H^{*}(-\boldsymbol{q}). \tag{10.42}$$

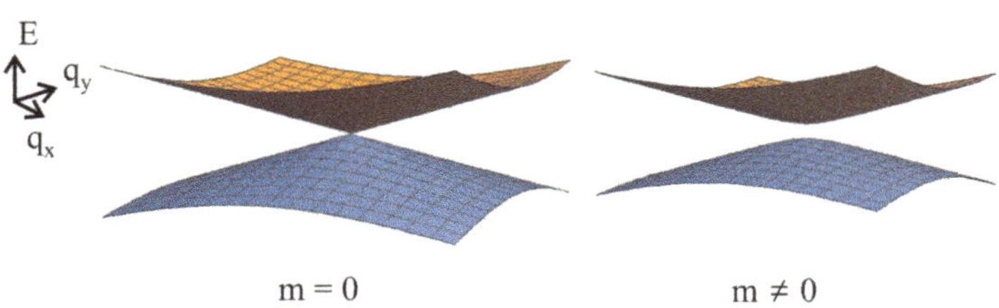

Figure 10.8. At left, a Dirac point exists at which the conduction and valence bands touch at an isolated point. At right: when time-reversal symmetry is broken, a mass term appears in the linearized Hamiltonian and an energy gap opens.

The combined PT operation therefore takes $H(\boldsymbol{q}) \to H^*(\boldsymbol{q})$. Terms proportional to σ_y, being imaginary, are absent from $H(\boldsymbol{q})$ since they would break the PT symmetry.

If either of the two symmetries is broken, then the gap opens at the Dirac point (right side of figure 10.8), which acts as a mass term in the Dirac equation. Taking $\hbar = c = 1$, the linearized Hamiltonian now takes the form $H(\boldsymbol{q}) = v(q_x\sigma_x + q_y\sigma_z) + m\sigma_y$, with dispersion relation $E(q) = \pm\sqrt{|v\boldsymbol{q}|^2 + m^2}$, leading to a mass gap of $\Delta E = 2m$ at $\boldsymbol{q} = 0$. The situation is different, though, depending on which symmetry is broken. If time reversal symmetry is broken (but inversion symmetry still holds), then the two Dirac points make equal contributions to the Berry curvature (each with a Berry flux of $+\pi$), leading to a Chern number of $c_1 = 1$. On the other hand, if it is the parity symmetry that is broken while time reversal invariance remains, then the two Dirac points have opposite sign curvature contributions, leading to a Chern number of $c_1 = 0$.

The Haldane model of graphene [14] is an example of a model where Dirac points can gain energy gaps. The Hamiltonian near each of the two Dirac points K and K' take the form given above, with mass terms arising due to symmetry breaking. If the inversion symmetry is broken, it is found that the excitations at the two Dirac points have equal mass, $m = m'$. In contrast, breaking of the time reversal symmetry leads to opposite sign masses: $m = -m'$.

In particle physics, a Dirac spinor is a solution to the relativistic Dirac equation and represents particles such as electrons. When the charge conjugation operator $\mathcal{C}$ is applied, the result is an antiparticle solution, $\psi_c = \mathcal{C}\psi$. In general, these particle and antiparticle solutions are distinct from each other. However, Ettore Majorana realized in 1937 that self-conjugate solutions are also possible that are invariant under charge conjugation. Such solutions represent particles which are equal to their own antiparticles. Although not seen so far as fundamental particles, in recent years it has been recognized that Majorana modes are likely to occur as collective excitations in condensed matter systems. Here, the roles of particle and antiparticle are played by electrons and holes.

Since condensed matter Majorana excitations must be invariant under particle–hole interchange, they must be formed as coherent superpositions of particle and hole states, similar to Cooper pairs. As a result, superconducting systems are a likely system in which to look for them, although superconductivity alone is not enough to guarantee their existence. It has been proposed that Majorana excitations may exist at the edges of some types of superconductors and of fractional quantum Hall systems.

More specifically, there is the possibility of **Majorana zero modes** in *topological* superconductors. These are zero-energy modes that exist in the middle of energy band gaps, and that remain localized near topological defects like vortices. What makes these Majorana zero modes especially interesting is that they can have fractional exchange statistics; in other words, rather than being bosons or fermions, they can act as non-Abelian anyons. (**Non-Abelian** means that the creation operators for these excitations don't commute with each other.) Anyons are of interest because they can form the basis for quantum computers, with computations being carried out by braiding the world-lines of anyons around each other [56], and then making measurements.

One additional type of spinor occurs in high-energy physics: a **Weyl spinor** is a spinor of definite chirality. (Chirality in this relativistic context is defined as the eigenvalue of a particular projection operator formed from the Dirac gamma matrices. At high energies it is the same as helicity, i.e. the projection of spin onto the direction of motion.) A Weyl spinor has half as many components as a Dirac spinor: a single Dirac spinor contains both chiralities, so each Dirac spinor can be thought of as being formed from a pair of opposite chirality Weyl spinors.

An analogous situation can occur in matter systems. A **Weyl point** is an isolated zero-energy point with linear dispersion in all three directions, so that the Hamiltonian has the form

$$H(\boldsymbol{q}) = v(q_x\sigma_x + q_y\sigma_y + q_z\sigma_z). \tag{10.43}$$

Because of the appearance of σ_y, these points only occur when PT symmetry is broken. The analog magnetic field due to the Berry connection is radial near the Weyl point, looking like the field from a magnetic monopole residing in momentum space, $\boldsymbol{B}(\boldsymbol{k}) = \pm\boldsymbol{k}/2|\boldsymbol{k}|^2$, with the sign being referred to as the **chirality** of the point. A Dirac point in three dimensions can be decomposed into a pair of Weyl points sitting at the same location, and it signals the degeneracy between four bands [57]. Mass terms in a Hamiltonian will always mix chiralities so they are necessarily absent from the Weyl Hamiltonian.

For more detailed discussion of Dirac points, see [23, 58]. See [59] for a review of Majorana fermions and Majorana zero modes, and [60] for Weyl semimetals.

References

[1] Fruchart M and Carpentier D 2013 *C. R. Phys.* **14** 779
[2] Carpentier D 2013 Seminaire Poincaré XVII 1
[3] Moore J E 2017 *An Introduction to Topological Phases of Electrons Topological Aspects of Condensed Matter Physics: Lecture Notes of the Les Houches Summer School* vol. 103 ed C Chamon, M O Goerbig, R Moessner and L F Cugliandolo (Oxford: Oxford University Press) p 3
[4] Hasan M Z and Kane C L 2010 *Rev. Mod. Phys.* **82** 3045
[5] Xiao D, Chang M-C and Niu Q 2010 *Rev. Mod. Phys.* **82** 1959
[6] Ashcroft N and Mermin D 1976 *Solid State Physics* (Boston, MA: Brooks Cole)
[7] Omar M A 1975 *Elementary Solid State Physics* (Reading, MA: Addison-Wesley)
[8] Griffiths D J 2017 *Introduction to Electrodynamics* 4th edn (Cambridge: Cambridge University Press)
[9] von Klitzing K, Dorda G and Pepper M 1980 *Phys. Rev. Lett.* **45** 494
[10] Paalanen M A, Tsui D C and Gossard A C 1982 *Phys. Rev.* B **25** 5566
[11] Tsui D C, Stormer H L and Gossard A C 1982 *Phys. Rev. Lett.* **48** 1559
[12] Laughlin R B 1983 *Phys. Rev. Lett.* **50** 1395
[13] Thouless D, Kohomoto M, Nightingale M and den Nijs M 1982 *Phys. Rev. Lett.* **49** 405
[14] Haldane F D M 1988 *Phys. Rev. Lett.* **61** 2015
[15] Chang C Z *et al* 2013 *Science* **340** 167

[16] Kane C L and Mele E J 2005 *Phys. Rev. Lett.* **95** 226081
[17] Maciejko J, Hughes T L and Zhang S C 2011 *Ann. Rev. Cond. Mat. Phys.* **2** 31
[18] Qi X L and Zhang S C 2010 *Phys. Today* **63** 33
[19] Girvin S M 2000 *Topological Aspects of Low Dimensional Systems* ed A Comtet, T Jolicoeur, S Ouvry and F David (Berlin: Springer)
[20] Taylor P L and Heinonen O 2002 *A Quantum Approach to Condensed Matter Physics* (Cambridge: Cambridge University Press)
[21] Goerbig M O 2009 arXiv:0909.1998v2 [cond-mat.mes-hall]
[22] Tong D 2016 arXiv:1606.06687 [hep-th]
[23] Stanescu T D 2017 *Introduction to Topological Quantum Matter and Quantum Computation* (Boca Raton, FL: CRC Press)
[24] Bernevig B A and (with Hughes T L) 2013 *Topological Insulators and Topological Superconductors* (Princeton, NJ: Princeton University Press)
[25] Asbóth J K, Oroszlány L and Pályi A P 2017 *A Short Course on Topological Insulators: Band Structure and Edge States in One and Two Dimensions* (Heidelberg: Springer)
[26] Su W P, Schrieffer J R and Heeger A J 1979 *Phys. Rev. Lett.* **42** 1698
[27] Su W P, Schrieffer J R and Heeger A J 1983 *Phys. Rev. Lett.* **22** 2099
[28] Batra N and Sheet G 2020 *Resonance J. Sci. Edu.* **25** 765
[29] Jackiw R and Rebbi C 1976 *Phys. Rev.* D **13** 3398
[30] Schnyder A P, Ryu S, Furusaki A and Ludwig A W W 2008 *Phys. Rev.* B **78** 195125
[31] Kitaev A 2009 *AIP Conf. Proc.* **1134** 22
[32] Chiu C-K, Teo J C Y, Schnyder A P and Ryu S 2016 *Rev. Mod. Phys.* **88** 035005
[33] Roy R and Harper F 2017 *Phys. Rev.* B **96** 155118
[34] Kohmoto M 1985 *Ann. Phys. (N.Y.)* **160** 343
[35] Wen X G 1995 *Adv. Phys.* **44** 405
[36] Read N and Green D 2000 *Phys. Rev.* B **61** 10267
[37] Bernevig B A, Hughes T L and Zhang S-C 2006 *Science* **314** 1757
[38] König M, Wiedmann S, Brüne C, Roth A, Buhmann H, Molenkamp L W, Qi X-L and Zhang S-C 2007 *Science* **318** 766
[39] Fu L, Kane C L and Mele E J 2007 *Phys. Rev. Lett.* **98** 106803
[40] Moore J E and Balintz L 2009 *Phys. Rev.* B **75** 195322
[41] Roy R 2007 *Phys. Rev.* B **79** 121306
[42] Fang K, Yu Z and Fan S 2012 *Nat. Photon.* **6** 782
[43] Rechtsman M C, Zeuner J M, Plotnik Y, Lumer Y, Podolsky D, Dreisow F, Nolte S, Segev M and Szameit A 2013 *Nature* **496** 196
[44] Hafezi J F A M M, Mittal S and Taylor J 2013 *Nat. Photon.* **7** 1001
[45] Maczewsky L J, Zeuner J M, Nolte S and Szameit A 2017 *Nat. Commun.* **8** 13756
[46] Mukherjee S, Spracklen A, Valiente M, Andersson E, Öhberg P, Goldman N and Thomson R R 2017 *Nat. Commun.* **8** 13918
[47] Teo J C Y, Fu L and Kane C L 2008 *Phys. Rev.* B **78** 045426
[48] Fu L 2011 *Phys. Rev. Lett.* **106** 106802
[49] Hsieh T H, Lin H, Liu J, Duan W, Bansil A and Fu L 2012 *Nat. Commun.* **3** 982
[50] Zhang L, Yang Y, Qin P, Chen Q, Gao F, Li E, Jiang J-H, Zhang B and Chen H 2020 *Adv. Science* **7** 1902724
[51] Hassan A E, Kunst F K, Moritz A, Andler G, Bergholtz E J and Bourennane M 2019 *Nat. Photon.* **13** 697

[52] Ota Y, Liu F, Katsumi R, Watanabe K, Wakabayashi K, Arakawa Y and Iwamoto S 2019 *Optica* **6** 786

[53] Xie B Y, Wang H F, Zhu X Y, Lu M H and Chen Y F 2018 *Phys. Rev.* B **98** 205147

[54] Yang Z, Lustig E, Lumer Y and Segev M 2020 *Light Sci. Appl.* **9** 128

[55] Nielsen H B and Ninomiya M 1981 *Phys. Lett.* B **105** 219

[56] Pachos J K 2012 *Introduction to Topological Quantum Computation* (Cambridge: Cambridge University Press)

[57] Young S M, Zaheer S, Teo J C Y, Kane C L, Mele E J and Rappe A M 2012 *Phys. Rev. Lett.* **108** 140405

[58] Pires A S T 2019 *A Brief Introduction to Topology and Differential Geometry in Condensed Matter Physics* (Bristol: IOP Publishing)

[59] Leijnse C and Flensberg K 2012 *Semicond. Sci. Tech.* **27** 124003

[60] Burkov A A 2016 *Nat. Mater.* **15** 1145

Chapter 11

Topological photonics

11.1 Overview: topological effects in photonic sytems

In the last chapter we discussed a range of topological effects that related to the band structure of electrons in solids. Analogs of many of these effects have recently been shown to appear in optical and photonic systems as well [1–6], beginning with the proposal of Haldane and Raghu [7, 8] to create an analog of the anomalous quantum Hall effect in optical systems. Optical systems have been found to be capable of supporting unidirectional topological edge states, gapped photonic bands with nonzero Chern number, and photonic analogues of the spin quantum Hall effect [9, 10], for example. Many applications of topological photonic states have been proposed, some of which will be described in the following sections.

Photonic systems obviously have significant differences from electronic systems, introducing hurdles that needed to be overcome. For example, (i) the Hall effect and many other topologically based effects require breaking of time reversal symmetry. Since photons have no charge, this cannot be done with magnetic fields as is it is in electronic systems. (ii) Another significant difference is the bosonic nature of photons, leading to the tendency of photons to crowd into the lowest available energy level instead of dividing up into distinct, well-separated bands. (iii) As seen in the last chapter, time-reversal symmetry acts differently on bosons and fermions. (iv) Photons, unlike electrons, do not interact directly with each other, although they can be made to interact indirectly, via nonlinear interactions mediated by a crystal lattice. (v) In addition, photons tend to propagate at the speed of light, meaning that any manipulation of their properties either has to be carried out very rapidly or has to be spread out spatially over the propagation path. (vi) Further, photons tend to be absorbed or scattered out of the system, leading to the need to be constantly pumping new photons in.

That last problem can actually be turned into an advantage. Loss or gain in an optical system is a non-Hermitian process, and it turns out that non-Hermitian processes offer new avenues for topological effects [11–19], including topological

doi:10.1088/978-0-7503-3471-6ch11

lasers (section 11.4). Photonic systems also provide a number of other advantages over electronic systems, such as the ability to more easily alter the system parameters in an experiment than is usually the case in conventional solid-state systems. Further, photonic systems do not require extreme cooling to observe topological effects, as is often the case in solid state or atomic systems.

The existence of the quantum Hall effect depends on the breaking of time reversal symmetry by the presence of a magnetic field. It was soon found that the presence or absence of particle-hole symmetry and of chiral symmetry (symmetry under interchange of two distinct types of lattice sites) are also relevant to the existence of topological phases. A classification of topological states possible with or without each of these symmetries has been constructed and is often referred to as the periodic table of topological insulators (chapter 10). Various mechanisms have been used to create symmetry conditions that allow the existence of topological states in photonics; for example, nonlinear devices called Faraday rotators can be used to break time reversal symmetry [20], the light can be sent into bipartite optical systems in which two distinct types of optical components alternate to create a chiral symmetry, or the system can be designed with some form of periodic driving. These approaches allow topological states to appear, for example, in cold atom optical lattices, coupled resonant oscillator systems, and photonic quantum walk systems.

In the next section we first look at topological effects induced by simple quantum walk systems for light. Then we look at effects arising in systems of waveguides, optical resonators, and dielectric photonic crystals. The field of topological photonics is expanding rapidly, so the following brief survey will necessarily be incomplete. The goal here is to give a brief guide to major threads of current research, rather than a detailed discussion of each.

It should be noted that, although we are discussing quantum systems here for the most part, some of these topological effects can also appear in classical optical networks [21].

11.2 Photonic walks

Quantum walks of photons on discrete lattices [22] are attractive because of their relative simplicity in experimental terms and the wide variety of phenomena that can be produced. Here, we only consider discrete quantum walks, in which a discrete step is taken at times $t = NT$, for some fixed T.

In a **classical random walk** [23], a particle lives on some discrete lattice (which we will assume to be one-dimensional for simplicity) and at each multiple of some discrete time T the particle jumps either left or right from its current location to one of the adjacent sites. The probability to jump right is some fixed value p, while the probability to jump left is $q = 1 - p$. As shown in every introductory statistical mechanics text [24–26], the probability of being at lattice site m after N time steps is given by a binomial distribution (see figure 11.1(a)),

$$P(m) = \frac{N!}{\left(\frac{N+m}{2}\right)!\left(\frac{N-m}{2}\right)!}p^{\frac{N+m}{2}}q^{\frac{N-m}{2}}. \tag{11.1}$$

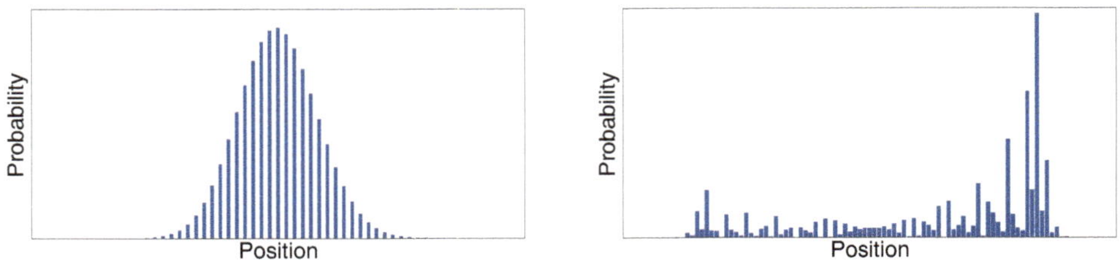

(a) (b)

Figure 11.1. (a) Classical random walk. The particle starts in the middle and has fixed probabilities to go left or right at each step. After N steps the probability distribution for positions is binomial and (for large N) approximately Gaussian. The spread is diffusive over time, $\sim\sqrt{N}$. The gaps at every other position are due to the fact that after an even number of steps only even positions can be reached (and similarly for odd positions after odd numbers of steps). (b) Quantum walk. The spread is now ballistic in time, $\sim N$. The asymmetry in the distribution is due to the choice of coin variable used; more symmetric walks can be produced with other coins or with different initial conditions.

When N is large, this becomes approximately Gaussian, as required by the central limit theorem. The width of the distribution as measured by the standard deviation grows proportionally to the square root of the time,

$$\sigma \sim \sqrt{N}. \tag{11.2}$$

This square root dependence on time is characteristic of diffusion processes, and so is called **diffusive spread**.

The behavior of **quantum walks** is very different [27–30]. In a quantum walk, there is a wavefunction instead of a classical particle. The wavefunction has both an amplitude to jump left *and* to jump right at each step, so it spreads in *both* directions with each step. After a few steps, there are multiple paths that can be taken from the initial point to other nearby lattice sites, resulting in quantum interference of different potential paths. The amplitudes for going left or right can be determined by the value of a random variable called the **coin variable**. The coin variable is a two-component vector; the components control the amplitude for leftward or rightward steps. The coin is rotated at each step, so that the amplitude for left and right steps varies. One way to describe the process is to use the **Feynman path integral approach** [31], which arrives at quantum amplitudes by summing over ensembles of classical trajectories. In any case, the amplitudes $a(m)$ for the particle to be at each site evolve deterministically over time and interfere with each other in a complicated manner, so that the final probability distribution is much more complex than in the classical case. The probability of being at site m at time N is then given by the absolute square of the probability amplitude, $P_N(m) = |a(m, N)|^2$. The probability distribution for a typical example is shown in figure 11.1(b). One difference is immediately apparent: the distribution is clearly non-Gaussian; in fact the probability tends to be small near the middle and largest at the edges. This is reflected in the behavior of the standard deviation over time: the width after N steps is now given by

$$\sigma \sim N. \tag{11.3}$$

The probability maxima near the edges move linearly in time like a freely propagating classical particle shot from a gun. For this reason, this behavior is called **ballistic spread**.

Quantum walks are of great interest in quantum information processing and quantum computing in part because of that last fact: since the quantum walk spreads faster than any classical random walk, the quantum walk can provide physical implementations of rapid quantum search algorithms. In fact, quantum walk processes can be used to model universal quantum computers [32–34]. A common feature of many quantum computation algorithms is that the superposition principle, interference, and other quantum properties lead to so-called **quantum speed-up** of algorithms [35, 36]. Although we won't go into detail here, holonomy and Berry phase play important roles in quantum walks; a discussion of this can be found in [37].

In a discrete quantum walk, what is actually relevant is the unitary time operator U that takes states forward one step in time, $|\psi(t)\rangle \to |\psi(t+T)\rangle = U|\psi(t)\rangle$. In this case we often define an effective Hamiltonian for the walk; since time-evolution operators are generally of the form $U(t) = e^{-iHt/\hbar}$, the effective Hamiltonian for the walk is defined to be

$$H = \frac{i\hbar}{T} \ln U, \tag{11.4}$$

where logarithms and exponentials of operators are defined as usual by their power series expansions. In this case, the U and H define a Floquet system (see the previous chapter), in which the discrete time step T is the driving period.

It has been demonstrated experimentally [38, 39] that one-dimensional quantum walks of photons in an optical system can simulate the behavior of Hamiltonians with non-trivial winding number, and that by bringing two such systems of different winding number together, topologically protected optical boundary states can be formed. This has been done using two-dimensional arrays of beam splitters and phase plates, with the photon moving along one axis and exhibiting a one-dimensional quantum walk along the perpendicular axis (figure 11.2). This is an example of a Floquet system, in which spatial variations of the system in the z direction

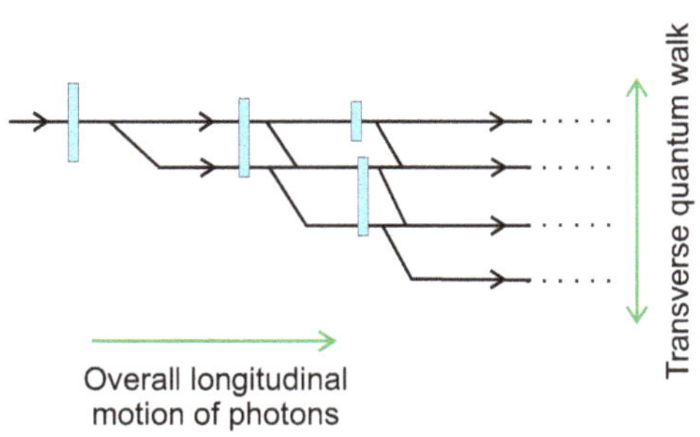

Figure 11.2. An optical implementation of a quantum walk. The random coin operation is carried out by rotating polarization states using half-wave plates (blue rectangles), then separating polarization components using birefringent elements (at the splitting points of the lines). The rotation sizes alternate to create nontrivial topological phases. The photons move left to right, but exhibit a quantum walk in the vertical direction.

provide the modulation. The variation in this case consists of using a so-called split-step walk, in which the each step of the walk consists of two separate substeps with different translation and coin rotation protocols; see [38, 39] for details. Another scheme which allows similar effects in a one-dimensional optical setup has also been proposed [40].

11.3 Photonic crystals, waveguides, and coupled resonant cavities

In 2005, Haldane and Raghu [7, 8] pointed out that an analog of the anomalous quantum Hall effect can be produced in optical systems. The system they discussed was a photonic crystal, in particular a gyromagnetic photonic crystal in which magneto-optical effects are used to break time reversal symmetry.

A **photonic crystal** [41], is an optically transparent material in which the electric permittivity and magnetic permeability are periodic functions of position. Examples include materials in which the refractive index is periodically modulated in one-, two-, or three-dimensions. Examples include alternating films of different index, a material with a periodic array of holes in it, or structures made from periodic arrangements of dielectric nanopillars (figure 11.3).

Changes in refractive index always lead to reflection of optical waves. So the periodic modulation in photonic crystals leads to interference between the incident and reflected waves. At some frequencies and wavenumbers, the interference will be constructive. Frequencies in this range will propagate in the crystal, providing allowed photonic bands analogous to electronic bands of a crystal. The ranges of frequency at which destructive interference occurs are forbidden bands; these frequency ranges will not propagate. These forbidden bands provide the energy gaps required for nontrivial topological states.

If a periodic array of high-refractive index structures is embedded in a lower-index medium, then the electric and magnetic field amplitudes tend to concentrate inside the high-index structures. So a collection of dielectric nanopillars or nanodots

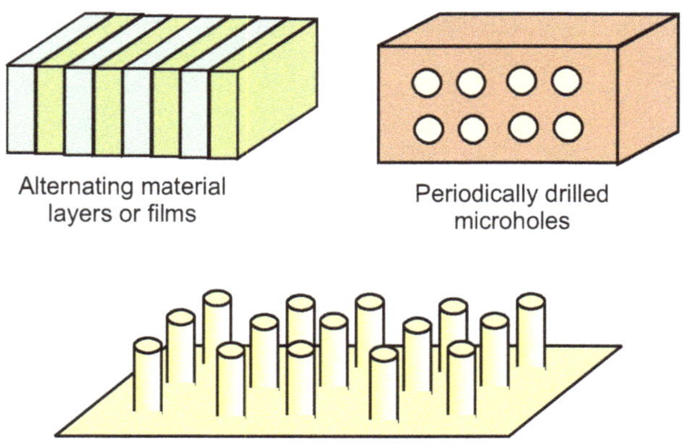

Figure 11.3. Examples of periodic structures in dielectric materials that can lead to the appearance of photonic bandgaps and allowed photonic frequency bands in photonic crystals.

can act as lattice sites where the field tends to localize, and by varying the refractive indices and the distances between structures, the hopping amplitudes of this localized field due to evanescent coupling between sites can be varied. In this way, common solid state structures and discrete hopping models such as a one-dimensional SSH model or a two-dimensional graphene-like honeycomb structure can be arranged.

It was predicted [7, 8] that two-dimensional photonic crystal systems with Hamiltonians of nontrivial Chern number could exist. These would display unidirectional edge states which would persist even in the presence of high levels of impurities or disorder; there would be no allowed states running in the other direction into which photons could scatter. When two of these systems in different topological phases are brought together, there would be boundary states trapped between them along the interface between different Chern numbers. These predictions were soon verified experimentally [42, 43]. Topologically protected states were first seen in microwave systems [42, 43], since this is the spectral region where the required magneto-optic effect is strong, but have now been detected in a variety of frequency ranges. The first topological insulator without the need for weak gyromagnetic effects was demonstrated experimentally in [44] using an array of evanescently coupled waveguides arranged in a graphene-like honeycomb pattern. The light travels monotonically along the z-axis, so that the z coordinate plays the role of an effective time variable. The waveguides wind in a helical shape in order to break the effective time-reversal symmetry.

Similar topologically based effects have since been demonstrated in a range of other systems [9, 10, 45–48], including optical cavities, quasicrystals, metamaterials, and lattices of coupled optical oscillators. At about the same time as the experiment of [44], topological insulators were also demonstrated in coupled systems of optical resonators [10, 49].

Optical resonators [50] are resonant cavities, often in the form of rings, in which optical fields can persist for long periods in resonant standing wave or circulating wave configurations. These can be coupled to each other or to waveguides by evanescant coupling; if two cavities are placed close enough together then the evanescent field from one cavity will extend a small distance into the other, allowing for tunneling of the field into the second cavity. This provides an inter-cavity coupling that can be engineered to have desired values with very high precision. Ring resonators have the added advantage that they can have optical excitations that circulate either clockwise or counter-clockwise, allowing simulation of other two-state systems like electron spin.

Figure 11.4(a) shows a pair of resonators (A and B) coupled to each other and to external waveguides. The evanescent coupling introduces a hopping phase ϕ. Notice that in this configuration the phases in each direction are equal, up to sign. The particular configuration shown in figure 11.4(b) adds an intermediate resonator (C) to introduce additional level of versatility and control. When C is shifted in the z direction, the two linkages between A and B are of different lengths. This provides an asymmetry between the left-moving and right-moving hopping phases for amplitudes. In this manner, the phase gained when travelling around a

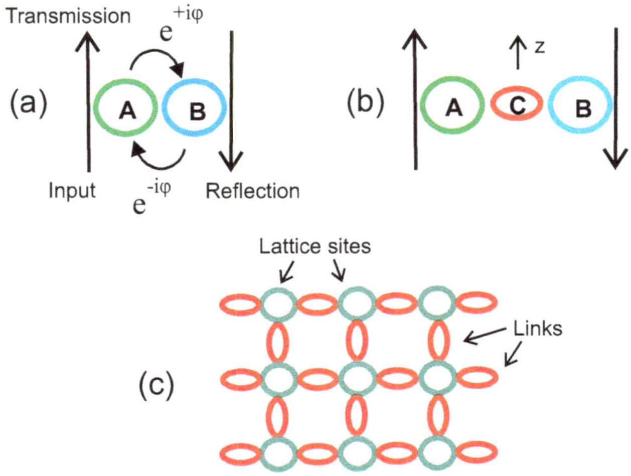

Figure 11.4. (a) Two optical ring resonators coupled evanescently to each other and to external waveguides. (b) A third resonator is used to couple the original two in order to provide additional functionality. (c) A two-dimensional array of coupled waveguides. The round resonators act as lattice sites, while the elliptical ones serve as links that provide hopping values that can be engineered to simulate different lattice models. Waveguides can be coupled to any of the lattice sites to provide input and output.

closed loop $e^{i\phi} = e^{-ie/hc\oint A\cdot dl}$ can simulate a gauge field (a *synthetic gauge field*) $\boldsymbol{A}$, which provides another means to break time reversal symmetry. The staggering of A/B resonators with C resonators also allows the imposition or breaking of chiral symmetry. A range of different possible topological systems can therefore be engineered.

A common two-dimensional configuration of coupled resonators is shown in figure 11.4(c). By engineering the hopping amplitudes provided by the intermediate links (the elliptical resonators), a wide variety of solid state lattice models can be implemented optically. In particular, it can be arranged so that the phases provided by the horizontal links can be varied from one row to the next. In this way, the phases around the upper and lower halves of a closed loop do not cancel, which provides a simulated gauge field with nonzero fluxes inside the areas enclosed by the loops.

11.4 Topologically protected waveguides and topological lasers

Recall that the interfaces between material regions with different winding number or Chern number support edge or boundary states that must remain confined to the vicinity of the boundary. These states are topologically protected: defects, impurities, or bendings in the boundary do not disrupt the states, and the states do not dissipate into the bulk regions. In addition, these states are generally chiral: they move in a single direction, and this direction is stable due to the lack of opposite-direction states in the same topological phase available for the particle to scatter into.

Such topological edge states can be useful in photonics for controlled transport of light without loss or scattering. In particular, high-coherence optical quantum states are often fragile and easily disrupted by interactions with the environment. These disruptions can decohere wavefunctions, meaning that the relative phases in quantum superpositions become randomized. This causes quantum effects such as interference and entanglement to be reduced or destroyed. Topologically protected waveguides would therefore be especially useful for decoherence-free transport of optical quantum states. A one-dimensional topologically protected waveguide for optical states is depicted in (figure 11.5). Similarly, topologically protected states on two-dimensional boundary regions between three-dimensional bulks can serve a way to confine photons to a plane in order to simulate two-dimensional physics. Two-dimensional topologically protected surface states may also soon find use in precision optical sensing applications. Topologically protected versions of waveguide splitters, directional filters, signal switches, and other optical devices have also been proposed or implemented.

An important ingredient of quantum optics, quantum information processing, and other areas is entanglement, two-particle or multi-particle states that cannot be factored into a product of single-particle states. Entangled states contain correlations between the particles that are stronger than any classical correlation [51]. So it was significant when it was shown that waveguides could be designed in which entangled states and quantum correlations can be topologically preserved [52–55].

All of the devices above are passive devices that require no external energy to operate. A further step can be taken to consider *active* topological photonics [56] that use pumping of energy from an external source to amplify signals or produce nonlinear effects. The chief example of active topological photonics is the **topological insulator laser** [57, 58]. The idea here is to start with an array of coupled ring oscillators and to pump only the resonators on the boundary of the system. Topological protection suppresses loss of the signal into the bulk, and prevents imperfections from disrupting the energy propagation on the boundary (figure 11.6). This was demonstrated in [57, 58], resulting in a high-coherence single-mode laser with higher ratios of laser intensity to pump intensity and higher output coupling efficiency than comparable topologically trivial lasers. The output was peaked near 1550 nm, in the wavelength window used for standard telecom applications. The bulk of the lasing action occurs at the boundary of the system. Defects in the boundary have little effect on the output of the topological laser; the energy simply

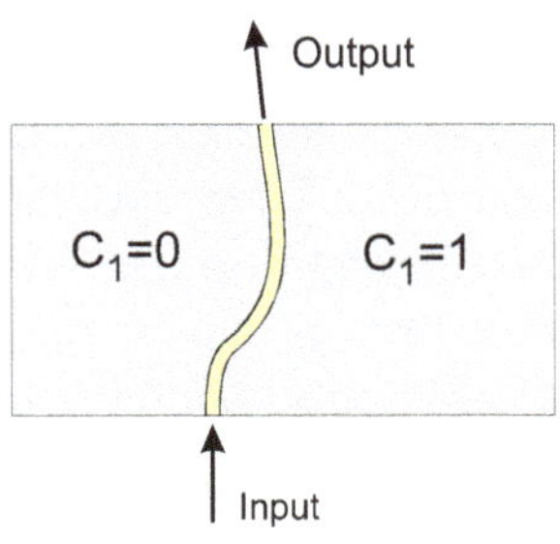

Figure 11.5. Two materials of different topological phases (different Chern numbers) are brought into contact, producing a state localized on the boundary (the yellow curve). The boundary can be used as a waveguide for optical states, with the topological protection serving to keep the state from being degraded by environmental factors.

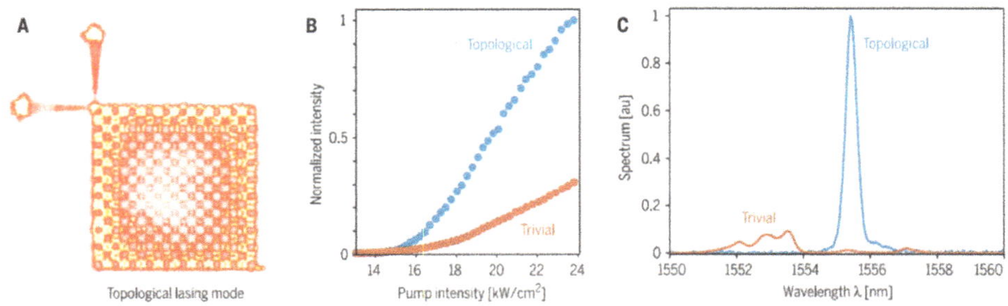

Figure 11.6. A topological insulator laser. (a) Resonator array pumped at periphery. The output port is at the upper left. Note that the amplitude remains near the boundary without leakage into the bulk, leading to higher-efficiency coupling to the output. (b) The single-mode output intensity compared to the maximum-output mode of a topologically trivial laser is improved by a factor of about 10. (c) The single-mode topological laser output is substantially narrower in bandwidth than the multi-mode trivial laser output. (Figures reproduced with permission from [58], copyright 2018 AAAS.)

detours a small distance into the bulk in order to bypass the defect without scattering, then returns to the boundary. This is in contrast to a nontopological laser, in which scattering at the defect disrupts laser emission in its vicinity.

A topological insulator laser was also demonstrated using gyromagnetic photonic crystals in [59], followed by a series of other variations [60–63], ranging from to topological quantum cascade lasers to topological vertical cavity surface emitting lasers. As a result of their topological stability, topological lasers can made with arbitrary shapes, and their performance is largely immune to imperfections in the fabrication process.

The surface of active topological photonics has barely been scratched. The future holds promise for further novel applications in sensing, antennas, and more exotic and specialized forms of topological lasers. In particular, nanocavity topological lasers [56, 64, 66] are in development which exhibit high speed and low power consumption, and which can be incorporated into integrated nanophotonic circuits.

11.5 Topological optical computing

One of the holy grails of physics in the early 21st century is the practical implementation of a fully programmable quantum computer. Such a computer would use purely quantum mechanical effects such as superposition and entanglement to carry out calculations orders of magnitude faster than classical computers. Although much progress has been made toward achieving this, the ultimate goal is still a ways off.

Many physical platforms have been proposed for quantum computing, ranging from cold atoms and MRI to spin systems [67], but a number of optical systems have long remained as promising candidates. Optical systems have advantages in versatility and practicality: they have many parameter values that can be easily tuned, are often easy to reconfigure, and do not need to be cooled to low temperatures. Regardless of the physical platform on which the computer is implemented, though, one major problem is that of decoherence and environmental disruption. Recall that quantum states obey the superposition principle: if ψ_1 and ψ_2 are solutions to the Schrödinger equation, then any linear combination is a solution as well. Any such superposition can always be written in the form

$$\psi = A(\psi_1 + B\, e^{i\phi}\psi_2), \tag{11.5}$$

where A is some (in general complex) constant, while B and ϕ are real constants. Many quantum effects, and in particular quantum computing effects, rely on the existence and stability of such superposition states. These effects are usually produced through interference of the terms in the superposition:

$$|\psi|^2 = |A|^2\left(|\psi_1|^2 + B^2|\psi_2|^2 + 2B\, \Re(e^{i\phi}\psi_1^*\psi_2)\right), \tag{11.6}$$

where the last term on the right is the interference term. ($\Re$ represents the real part.) An entangled state is simply a superposition state in which each term in the superposition is a multi-particle state.

These quantum states and the ability to produce interference patterns from them are very fragile. In particular, interactions with the environment often cause the relative phase $e^{i\phi}$ to fluctuate randomly. Unless carefully controlled, this random fluctuation will cause the interference term to oscillate wildly, so that any attempt to measure the interference term will average to zero over the finite times required for real measurements. The interference term will be washed out by the oscillation, and the desired quantum effects will become unobservable. This process, called **decoherence**, leads the quantum system to behave classically and destroys any attempt at quantum computation. There are many ways proposed to get around this decoherence, for example by isolation from the environment and use of low temperatures, or by building redundancy into the system. But one intriguing possibility is to use topology.

The idea behind **topological quantum computing** [68, 69] is to encode information into a topological invariant, and to take advantage of topological protection to make the quantum information stable against decoherence and other disturbances. Probably the first proposal along these lines was to use collective anyonic excitations of spins on a surface to carry out fault-tolerant computations [70]. Other potential topologically stable carriers for quantum information include the use of fractional quantum Hall states and Majorana zero modes.

In the optical regime, we have seen (chapter 6) that OAM states of light have vortices at their center with topologically stable angular momentum quantum numbers, and that in principle very high angular momenta can be achieved, allowing multiple bits of information to be transmitted with a single photon. But we also saw that high angular momentum states tend to split into multiple lower-OAM vortices and that these vortices can wander far apart from each other, making the total quantum number of high OAM states difficult to measure accurately. Similarly, we saw that knotted or linked optical vortices could be used to store information in topological invariants (chapter 7), but it is difficult to see how the information can be encoded, processed to carry out a computation, and then read back out in a practical manner in the near future.

Cluster states [71, 72] are a type of highly entangled multi-particle quantum state that are currently the subject of much interest. Computations can be done by performing a sequence of measurements on the cluster in a particular order. Optical cluster states have been discussed as a means of carrying out topological quantum computation [73–75]. These states have the advantage that they can be scaled up to large sizes with many quantum bits (qubits) involved in the computation.

It has been long been known that braiding of anyons [76], if it could be accomplished, could be used to perform topologically protected quantum computations. When two identical particles in three dimensions are interchanged, their total wavefunction always gains a phase factor of $+1$ for bosons or -1 for fermions. The spin-statistics theorem guarantees that in three dimensions these are the only possibilities. But in two dimensions anyons, particles or quasiparticles of fractional statistics, can occur which can gain arbitrary phases $e^{i\phi}$ under interchange. An array of multiple anyons can be interchanged sequentially, so that their world lines are braided. Each interchange carries its own phase, and these phases form a group, the **braid group**. (See the appendix for the definition of a group.) In the simplest case where the states are nondegenerate (so each configuration of particles corresponds to a unique quantum state), the group is Abelian, meaning that the braiding operations commute: operations carried out in different orders produce the same result.

However, the system can be degenerate, so that the same configuration of particles can correspond to multiple distinct quantum states. The Berry connections are now matrix-valued, and these matrices do not necessarily commute. These non-Abelian or non-commutative anyons can then form the basis for quantum computation, with computations being carried out by braiding followed by measurement [68, 69].

A mechanism for non-Abelian braiding of light was proposed in reference [77], and was implemented experimentally in [78]. The idea is to create a photonic crystal with topological defects that braid around each other. The defects act as light guides for optical vortices which then pick up non-Abelian Berry phases as their positions are interchanged. The non-Abelian nature of the braiding is verified by splitting the initial beam into two copies, sending them into photonic crystals with the braids occurring in different orders, then recombining the beams on a beam splitter. The resulting interference pattern demonstrates that the two different braid orderings produce different phases. This non-Abelian interferometer is shown schematically in figure 11.7. It should be noted that the non-Abelian interference occurs at both the quantum (single particle) and classical (coherent state) levels.

Finally, rather than a full universal quantum computer that can carry out arbitrary computations, a simpler task is to build a quantum simulator, a special-purpose quantum computer that simulates some other physical phenomena. The working of an optical quantum simulator has been demonstrated [79] that simulated the time evolution of Majorana fermions under non-Abelian braiding.

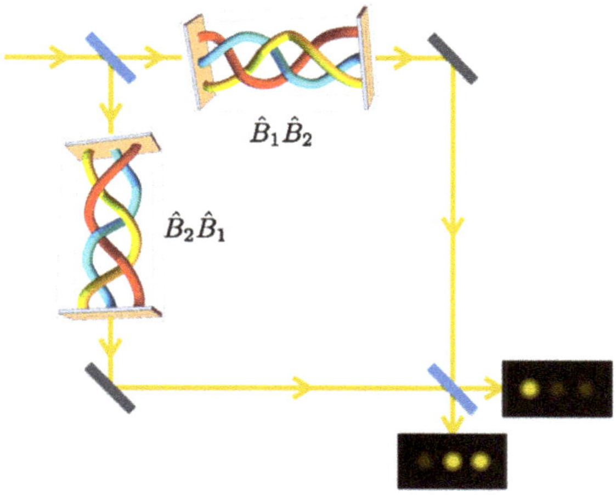

Figure 11.7. A schematic depiction of an interferometer demonstrating the non-Abelian braiding of light. Identical copies of the input undergo the same braid in different orders before being recombined at a beam splitter. The two complementary interference patterns at the final beam splitter output show that the different orderings lead to the accumulations of different Berry phases. (Figures reproduced with permission from [77], copyright 2015 American Physical Society.)

The fields of topological photonics and of topological quantum computing with light are still very much in their infancy, but have already shown substantial results. The potential for future developments is enormous and is bound to lead to surprises.

References

[1] Lu L, Joannopoulos J D and Soljačić M 2014 *Nat. Photonics* **8** 821
[2] Khanikaev A B and Shvets G 2017 *Nat. Photonics* **11** 763
[3] Kitagawa T 2012 *Quantum Inf. Process.* **11** 1107
[4] Ozawa T *et al* 2019 *Rev. Mod. Phys.* **91** 015006
[5] Hafezi M and Taylor J 2017 *Quantum Simulations with Photons and Polaritons* ed D G Angelakis (Berlin: Springer)
[6] Segev M and Bandres M A 2021 *Nanophotonics* **10** 425
[7] Haldane F D M and Raghu S 2008 *Phys. Rev. Lett.* **100** 013904
[8] Raghu S and Haldane F D M 2008 *Phys. Rev.* A **78** 033834
[9] Hafezi M, Demler E A, Lukin M D and Taylor J M 2011 *Nat. Phys.* **7** 907
[10] Hafezi M, Mittal S, Fan J, Migdall A and Taylor J M 2013 *Nat. Photonics* **7** 1001
[11] Zeuner J M, Rechtsman M C and Plotnik Y *et al* 2015 *Phys. Rev. Lett.* **115** 040402
[12] Longhi S 2017 *Europhys. Lett.* **120** 64001
[13] El-Ganainy R, Makris K G, Khajavikhan M, Musslimani Z H, Rotter S and Christodoulides D N 2018 *Nat. Phys.* **14** 11
[14] Rudner M S, Levin M and Levitov L S 2016 arXiv:1605.07652 Cond-Mat
[15] Lee T E 2016 *Phys. Rev. Lett.* **116** 133903
[16] Leykam D, Bliokh K Y, Huang C, Chong Y D and Nori F 2017 *Phys. Rev. Lett.* **118** 040401
[17] Shen H, Zhen B and Fu L 2018 *Phys. Rev. Lett.* **120** 146402

[18] Gong Z, Ashida Y, Kawabata K, Takasan K, Higashikawa S and Ueda M 2018 *Phys. Rev.* X **8** 031079
[19] Xiao L, Deng T and Wang K *et al* 2020 *Nat. Phys.* **16** 761
[20] Koch J, Houck A A, Hur K and Girvin S M 2010 *Phys. Rev.* A **82** 043811
[21] Shi T, Kimble H J and Cirac J I 2017 *Proc. Natl. Acad. Sci. USA* **114** E896
[22] Tarasinski B, Asbóth J K and Dahlhaus J P 2014 *Phys. Rev.* A **89** 042327
[23] Klafter J and Sokolov M 2011 *First Steps in Random Walks* (Oxford: Oxford University Press)
[24] Reif F 1965 *Fundamentals of Statistical and Thermal Physics* (Long Grove, IL: Waveland)
[25] Kittel C 1958 *Elementary Statistical Physics* (Mineola, NY: Dover)
[26] Gould H and Tobochnik J 2010 *Statistical and Thermal Physics: With Computer Applications* (Princeton, NJ: Princeton University Press)
[27] Aharonov, Davidovich L and Zagury N 1993 *Phys. Rev.* A **48** 1687
[28] Kempe J 2003 *Contemp. Phys.* **44** 307
[29] Venegas-Andraca S E 2012 *Quantum Inf. Process.* **11** 1015
[30] Manouchehri K and Wang J 2014 *Physical Implementation of Quantum Walks* (Berlin: Springer)
[31] Feynman R P and Hibbs A R 1965 *Quantum Mechanics and Path Integrals* (New York: McGraw-Hill)
[32] Ambainis A 2003 *Int. J. Quant. Inf.* **1** 507
[33] Portugal R 2013 *Quantum Walks and Search Algorithms* (Berlin: Springer)
[34] Venegas-Andraca S E 2008 *Quantum Walks for Computer Scientists* (San Rafael, CA: Morgan and Claypool)
[35] Nielsen M A and Chuang I L 2000 *Quantum Computation and Quantum Information* (Cambridge: Cambridge University Press)
[36] Mermin N D 2007 *Quantum Computer Science: An Introduction* (Cambridge: Cambridge University Press)
[37] Puentes G 2017 *Crystals* **7** 122
[38] Broome M A, Fedrizzi A, Lanyon B P, Kassal I, Aspuru-Guzik A and White A G 2010 *Phys. Rev. Lett.* **104** 153602
[39] Kitagawa T, Broome M, Fedrizzi A, Rudner M S, Berg E, Kassal I, Aspuru-Guzik A, Demler E and White A G 2012 *Nat. Comm.* **3** 882
[40] Simon D S, Fitzpatrick C A, Osawa S and Sergienko A V 2017 *Phys. Rev.* A **96** 013858
[41] Joannopoulos J, Johnson S, Winn J and Meade R 2008 *Photonic Crystals: Molding the Flow of Light* (Princeton, NJ: Princeton University Press)
[42] Wang Z, Chong Y D, Joannopoulos J D and Soljačić M 2008 *Phys. Rev. Lett.* **100** 013905
[43] Wang Z, Chong Y D, Joannopoulos J D and Soljačić M 2009 *Nature* **461** 772
[44] Rechtsman M C, Zeuner J M and Plotnik Y *et al* 2013 *Nature* **496** 196
[45] Cho J, Angelakis D G and Bose S 2008 *Phys. Rev. Lett.* **101** 246809
[46] Fang K, Yu Z and Fan S 2011 *Phys. Rev.* B **84** 075477
[47] Liu K, Shen L and He S 2012 *Opt. Lett.* **37** 4110
[48] Kraus Y E, Lahini Y, Ringel Z, Verbin M and Zilberberg O 2012 *Phys. Rev. Lett.* **109** 106402
[49] Mittal S, Fan J, Faez S, Migdall A, Taylor J M and Hafezi M 2014 *Phys. Rev. Lett.* **113** 087403

[50] Hodgson N and Weber H 2005 *Laser Resonators and Beam Propagation* 2nd edn (Berlin: Springer)

[51] Greenstein G and Zajonc A 2006 *The Quantum Challenge: Modern Research on the Foundations of Quantum Mechanics* (Sudbury, MA: Jones and Bartlett)

[52] Blanco-Redondo A, Bell B, Eggleton B J and Segev M 2018 *Science* **362** 568

[53] Wang M, Doyle C and Bell B *et al* 2019 *Nanophotonics* **8** 1327

[54] Mittal S, Orre V V and Hafezi M 2016 *Opt. Exp.* **24** 15631

[55] Rechtsman M C, Lumer Y, Plotnik Y, Perez-Leija A, Szameit A and Segev M 2016 *Optica* **3** 925

[56] Ota Y, Takata K and Ozawa T *et al* 2020 *Nanophotonics* **9** 547

[57] Harari G, Bandres M A and Lumer Y *et al* 2018 *Science* **359** eaar4005

[58] Bandres M A, Wittek S and Harari G *et al* 2018 *Science* **359** eaar4005

[59] Bahari B, Ndao A, Vallini F, El Amili A, Fainman Y and Kanté B 2017 *Science* **358** 636

[60] Zeng Y, Chattopadhyay U and Zhu B *et al* 2020 *Nature* **578** 246

[61] Amelio I and Carusotto I 2020 *Phys. Rev.* **101** 064505

[62] Shao Z-K, Chen H Z and Wang S *et al* 2020 *Nat. Nanotechnol.* **15** 67

[63] Liu Y G, Jung P, Parto M, Hayenga W E, Christodoulides D N and Khajavikhan M 2020 *Novel In-Plane Semiconductor Lasers XIX* ed A A Belyanin and P M Smowton (Bellingham, WA: SPIE) 36

[64] Han C, Lee M, Callard S, Seassal C and Jeon H 2019 *Light Sci. Appl.* **8** 40

[65] Ota Y, Katsumi R, Watanabe K, Iwamoto S and Arakawa Y 2018 *Commun. Phys.* **1** 86

[66] Pilozzi L and Conti C 2016 *Phys. Rev.* B **93** 195317

[67] Nielsen M A and Chuang I L 2010 *Quantum Computation and Quantum Information: 10th Anniversary Edition* (Cambridge: Cambridge University Press)

[68] Pachos J K 2012 *Introduction to Topological Quantum Computation* (Cambridge: Cambridge University Press)

[69] Nayak C, Simon S H, Stern A, Freedman M and Das Sarma S 2008 *Rev. Mod. Phys.* **80** 1083

[70] Kitaev A Y 2003 *Ann. Phys.* **303** 2

[71] Briegel H J and Raussendorf R 2001 *Phys. Rev. Lett.* **86** 910

[72] Nielsen M A 2006 *Rep. Math. Phys.* **57** 147

[73] Raussendorf R, Harrington J and Goyal K 2007 *New J. Phys.* **9** 199

[74] Devitt S 2010 *Optics and Spectroscopy* **108** 267

[75] Rudolph T 2017 *APL Photonics* **2** 030901

[76] Wilczek F and Zee A 1984 *Phys. Rev. Lett.* **52** 2111

[77] Iadecola T, Schuster T and Chamon C 2015 *Phys. Rev. Lett.* **117** 073901

[78] Noh J, Schuster T, Iadecola T, Huang S, Wang M, Chen K P, Chamon C and Rechstman M C 2020 *Nat. Phys.* **16** 989

[79] Xu J-S, Sun K, Han Y-J, Li C-F, Pachos J K and Guo G-C 2016 *Nat. Commun.* **7** 13194

IOP Publishing

Topology in Optics (Second Edition)

Tying light in knots

David S Simon

Appendix A

Appendices

A.1 Point-set topology: basic definitions and results

Topologically interesting spaces are often those that cannot be covered consistently by a single coordinate system that extends over the whole space. Examples are the Möbius strip, where functions (including coordinate values) can become double-valued upon multiple circuits around the strip, and the sphere, where a coordinate system with origin at one pole will have a singularity at the other pole (where the coordinate axes inevitably run into each other).

In cases such as these, the space must be covered with multiple coordinate patches, defined on overlapping open sets. On the overlaps, there must be transition functions defining the change from one coordinate system to the other. The information about the global structure of the space is then encoded into these transition functions. In physics, the transition functions between the open sets often correspond to gauge transformations.

Topology textbooks therefore begin with detailed studies of open sets and the formal definition of a topological space is given in terms of **charts**, collections of open sets that obey a set of consistency requirements. Here, we give a brief overview of these definitions and a few important results, stated without proofs. For more details, see any standard topology text, such as [1–4]

Consider a set X. A **topology** $\mathcal{T}$ on X is a collection of subsets U_i, where $i = 1, 2, \ldots$, such that:

(i) $\mathcal{T}$ contain both the empty set $\varnothing$ and the full original set X: $\varnothing, X \in \mathcal{T}$

(ii) $\mathcal{T}$ is closed under finite or countably infinite unions: $\cup_{i\in\Lambda} U_i \in \mathcal{T}$, with Λ being a finite or infinite discrete set of labels

(iii) $\mathcal{T}$ is closed under finite intersections: $\cup U_{i\in\Lambda} \in \mathcal{T}$, with Λ being a finite discrete set of labels

The pair $(X, \mathcal{T})$ is a **topological space**. For brevity, X is commonly referred to by itself as a topological space, but it is understood that a topology $\mathcal{T}$ must also be provided. The collection of sets $U_i \in \mathcal{T}$ are called the **open sets** of the topology.

doi:10.1088/978-0-7503-3471-6ch12

Examples.

(1) The **usual topology** on $\mathbb{R}$: $\mathcal{T}$ is defined to be the collection of all open intervals (a, b) in $\mathbb{R}$, along with their unions. To go to higher dimensions, the usual topology on $\mathbb{R}^n$ is then defined by letting the open sets on $\mathbb{R}^n$ be the Cartesian products of the open sets on $\mathbb{R}$.

(2) The **metric topology**. Let X be a metric space with distance function $d(x, y)$. Then the open sets of $\mathcal{T}$ can be taken to be the set of balls or disks about each point in X, i.e. the set of points y such that $d(x_0, y) < r$ for each point $x_0 \in X$ and for all radii r.

A **closed set** in X is the complement of an open set: U is open if $\bar{U} = X - U$ is open. In addition to open and closed sets, it is often useful to define neighborhoods of each point: $\mathcal{N}$ is a **neighborhood** of point $x_0 \in X$ if $\mathcal{N}$ contains at least one open set U of X that in turn contains x_0: $x_0 \in U \subset \mathcal{N}$. The neighborhood itself need not be open. X is **connected** if it is not the union of two disjoint open sets of X; in other words, there are no open subsets $U_1, U_2 \in X$ such that $X = U_1 \cup U_2$ and $U_1 \cap U_2 = \varnothing$.

The **interior** $int(Y)$ of subset $Y \in X$ is the largest open subset contained in Y. The **boundary** ∂Y of Y is the complement of the interior: $\partial Y = Y - int(Y)$. A **covering** $\mathcal{C}$ of X is a family of subsets S_j (not necessarily open) of X whose union covers all of X: $\cup_j S_j = X$. Covers of a space are not unique. A **subcovering** $\mathcal{C}'$ of $\mathcal{C}$ is a subcollection of the sets in the original cover $\mathcal{C}$, which form a covering of X themselves. X is called **compact** if every cover of X has a finite subcover (a subcover containing a finite number of sets S_j). It is often convenient to turn a noncompact set into a compact one by adding a point to it. For example, the noncompact set $\mathbb{R}$ can be made compact by adding a single point at infinity and identifying it with a similar point at $-\infty$; this turns the line $\mathbb{R}$ into a circle S^1. Similarly, the noncompact plane $\mathbb{R}^2$ is turned into a compact two-dimensional sphere S^2 by adding a point at infinity and identifying all points at infinite distance from the origin in the plane with this new point. This construction is referred to as forming the **one-point compactification** of the original noncompact set.

A map between two topological spaces, $f: X \to Y$, is called a **continuous map** if the inverse image $f^{-1}(U)$ of each open set $U \in Y$ is an open set in X. If the map is both continuous and invertible, then it is a **homeomorphism**. Two topological spaces connected by a homeomorphism are said to be **homeomorphic** to each other. **Topological invariants** are quantities that remain invariant under homeomorphisms.

Topology is largely about the classification of spaces by grouping them into categories, based on which spaces are homeomorphic to each other. Computing topological invariants provides an easy way of distinguishing between two inequivalent spaces: if the invariants have different values then the spaces are not homeomorphic. In addition to winding number and Chern number, examples of topological invariants include the genus (number of handles, g) and the closely related Euler characteristic, χ; these are all discussed in chapter 5.

A.2 Brief review of group theory

Group theory is the branch of abstract algebra that studies transformations and symmetries of mathematical objects. First formalized by Galois, Cayley, Cauchy, Lie, and others from the 1830s onward, it was initially used to study the solvability of algebraic equations. Over the course of the twentieth century it not only became an essential tool in other areas of mathematics, such as differential geometry and topology, but became indispensable throughout the natural sciences, especially physics and chemistry.

A group consists of a set $\mathcal{S}$ together with a binary operation '·' acting on pairs of elements in $\mathcal{S}$. The binary operation is usually referred to as group multiplication. The set and multiplication operation are required to satisfy the following conditions:

- **Closure:** If x and y are elements of $\mathcal{S}$, then $x \cdot y$ is also in $\mathcal{S}$.
- **Identity:** There exists a special element $\mathcal{I} \in \mathcal{S}$, called the identity element, such that $x \cdot \mathcal{I} = \mathcal{I} \cdot x = x$ for all $x \in \mathcal{S}$.
- **Inverses:** For every element $x \in \mathcal{S}$ there is another element (called the inverse of x and usually written as x^{-1}) such that $x \cdot x^{-1} = x^{-1} \cdot x = \mathcal{I}$.
- **Associativity:** For all elements x, y, and z in $\mathcal{S}$, we have $(x \cdot y) \cdot z = x \cdot (y \cdot z)$.

Note that the multiplication operation is not necessarily commutative: $x \cdot y$ may not equal $y \cdot x$. If the multiplication is commutative, then the group is called **Abelian**; otherwise, it is **non-Abelian**.

One of the simplest examples of a group is the set of integers, $\mathcal{S} = \mathbb{Z}$, with the usual addition operation playing the role of group multiplication, $\cdot \rightarrow +$. It is easy to verify that the group properties are all satisfied, with the integer 0 playing the role of the identity element, and inverses given by negatives: $x^{-1} = -x$.

More often in physics, the elements of $\mathcal{S}$ are the set of transformations that leave some object or mathematical structure invariant. For example, consider a square in the x-y plane. The square is invariant under a set of rotations, $\mathcal{S} = \{R(0), R(\pm\pi/2), R(\pm 2\pi/2), R(\pm 3\pi/2), \ldots\}$, where $R(\theta)$ is a rotation in the plane through angle θ. Thinking of these rotations as matrices, the group multiplication operation is then just the usual matrix multiplication. Note that reflections about the lines bisecting the sides of the square and reflections across the diagonals also leave the square invariant; these can be adjoined to the rotations to form a larger group. This larger group is the full symmetry group of the square, and the group of rotations forms a subgroup of it.

This illustrates the most important use of group theory in physics: groups describe *symmetries* of an object. Rather than a concrete object like a square, the group will often represent the symmetries of a more abstract object like a Lagrangian or a set of quantum fields. Symmetries play a fundamental role in physics for two related reasons: (i) Continuous symmetries imply the existence of conservation laws (a result known as **Noether's theorem**) and (ii) maintaining these symmetries forces the introduction of gauge fields (electromagnetic, nuclear, and gravitational fields). Symmetry principals are therefore major building blocks in the modern approach to physics.

The most important class of groups in physics are Lie groups; these are groups whose elements form a continuous, differential manifold, rather than a discrete set. The unitary groups $SU(n)$ are common examples of Lie groups and represent rotations in complex n-dimensional spaces. They are widely used in particle physics. The special case of $SU(2)$ also comes up in information theory and in optics, for example, in the description of beam splitters. The orthogonal groups $SO(n)$ represent rotations in real n-dimensional spaces.

The simplest Lie group is $U(1)$, which acts as the gauge group of electromagnetism. $U(1)$ is the multiplicative group of complex numbers with absolute value 1; in other words, its elements are the continuous set of phase factors, $g(\theta) = e^{i\theta}$. Ordinary multiplication of complex numbers serves as the group product, $g(\theta_1)g(\theta_2) = e^{i(\theta_1+\theta_2)} = g(\theta_1 + \theta_2)$. Topologically, the group is one dimensional and forms a circle S^1 in the complex plane. $U(1)$ is an Abelian group and is isomorphic to the group $SO(2)$ of rotations in a plane.

Lie groups, being differentiable manifolds, have tangent spaces at each point. This vector space and a binary operation (the commutator, $[a, b] = ab - ba$) are the basic ingredients in forming a Lie algebra. More information on Lie algebras and Lie groups may be found in many references, for example, [5–7].

The set of homotopy classes of a topological space also form groups, the homotopy groups π_n (see chapter 3 for details). For reference, we list some useful homotopy groups:

(I) **Spheres:**

Sphere	π_1	π_2	π_3	π_4	π_5	π_6
S^1	$\mathbb{Z}$	0	0	0	0	0
S^2	0	$\mathbb{Z}$	$\mathbb{Z}$	$\mathbb{Z}_2$	$\mathbb{Z}_2$	$\mathbb{Z}_{12}$
S^3	0	0	$\mathbb{Z}$	$\mathbb{Z}_2$	$\mathbb{Z}_2$	$\mathbb{Z}_{12}$
S^4	0	0	0	$\mathbb{Z}$	$\mathbb{Z}_2$	$\mathbb{Z}_2$

Here, $\mathbb{Z}_n$ means the group of integers mod(n). For example, $\mathbb{Z}_3$ is the three-element cyclic group $\{1, 2, 3\}$; the numbers 4, 7, 10, ... are all equivalent to the element 1, while 5, 8, 11, ... are equivalent to 2, and so on.

(II) ***n*-dimensional torus:** $\pi(T^n) = \mathbb{Z}^n$ (the product of n copies of $\mathbb{Z}$).

(III) **Unitary groups**

Group	π_1	π_2	π_3	π_4	π_5	π_6	π_7	π_8
$U(1)$	$\mathbb{Z}$	0	0	0	0	0	0	0
$SU(2)$	0	0	$\mathbb{Z}$	$\mathbb{Z}_2$	$\mathbb{Z}_2$	$\mathbb{Z}_{12}$	$\mathbb{Z}_2$	$\mathbb{Z}_2$
$SU(3)$	0	0	$\mathbb{Z}$	0	$\mathbb{Z}$	$\mathbb{Z}_6$	0	$\mathbb{Z}_{12}$
$SU(4)$	0	0	$\mathbb{Z}$	0	$\mathbb{Z}$	0	$\mathbb{Z}$	$\mathbb{Z}_{24}$
$SU(5)$	0	0	$\mathbb{Z}$	0	$\mathbb{Z}$	0	$\mathbb{Z}$	0

Bott Periodicity for $SU(n)$. For $k > 1$, $n \geqslant (k+1)/2$:

$$\pi_k(U(n)) = \pi_k(SU(n)) = \begin{cases} 0 & \text{for even } k \\ \mathbb{Z} & \text{for odd } k \end{cases} \tag{A.1}$$

Fundamental groups. For all n: $\pi_1(SU(n)) = 0$ and $\pi_1(U(n)) = 1$.

(IV) **Orthogonal groups**

Group	π_1	π_2	π_3	π_4	π_5	π_6	π_7	π_8
$SO(2)$	$\mathbb{Z}$	0	0	0	0	0	0	0
$SO(3)$	$\mathbb{Z}_2$	0	$\mathbb{Z}$	$\mathbb{Z}_2$	$\mathbb{Z}_2$	$\mathbb{Z}_{12}$	$\mathbb{Z}_2$	$\mathbb{Z}_2$
$SO(4)$	$\mathbb{Z}_2$	0	$\mathbb{Z}\times\mathbb{Z}$	$\mathbb{Z}_2\times\mathbb{Z}_2$	$\mathbb{Z}_2\times\mathbb{Z}_2$	$\mathbb{Z}_{12}\times\mathbb{Z}_{12}$	$\mathbb{Z}_2\times\mathbb{Z}_2$	$\mathbb{Z}_2\times\mathbb{Z}_2$
$SO(5)$	$\mathbb{Z}_2$	0	$\mathbb{Z}$	$\mathbb{Z}_2$	$\mathbb{Z}_2$	0	$\mathbb{Z}$	0
$SO(6)$	$\mathbb{Z}_2$	0	$\mathbb{Z}$	0	$\mathbb{Z}$	0	$\mathbb{Z}$	$\mathbb{Z}_{24}$

Bott Periodicity for $SO(n)$. For $n \geqslant k+2$:

$$\pi_k(O(n)) = \pi_k(SO(n)) = \begin{cases} 0 & \text{for } k = 2,4,5,6 (\text{mod } 8) \\ \mathbb{Z}_2 & \text{for } k = 0,1 (\text{mod } 8) \\ \mathbb{Z} & \text{for } k = 3,7 (\text{mod } 8) \end{cases} \tag{A.2}$$

References

[1] Hatcher A 2002 *Algebraic Topology* (Cambridge : Cambridge University Press)

[2] Greenberg M and Harper J 2018 *Algebraic Topology: A First Course* (Boca Raton, FL: CRC Press)

[3] Munkres J R 2000 *Topology* (Upper Saddle River, NJ: Prentice-Hall)

[4] Basener W F 2006 *Topology and Its Applications* (Hoboken, NJ: Wiley)

[5] Warner F W 1983 *Foundations of Differentiable Manifolds and Lie Groups* (Berlin: Springer)

[6] Gilmore R 2006 *Lie Groups, Lie Algebras, and Some of Their Applications* (Mineola, NY: Dover)

[7] Zee A 2016 *Group Theory in a Nutshell for Physicists* (Princeton, NJ: Princeton University Press)

www.ingramcontent.com/pod-product-compliance
Ingram Content Group UK Ltd.
Pitfield, Milton Keynes, MK11 3LW, UK
UKHW051851100926
8556IPUK00005B/569

9 780750 334